TP 492.7 . ⊘   S0-AZM-879

Wylie, David, 1953-

...igerants for air

New Refrigerants for Air Conditioning and Refrigeration Systems

New Refrigerants for Air Conditioning and Refrigeration Systems

David Wylie, P.E.

James W. Davenport

Published by
THE FAIRMONT PRESS, INC.
700 Indian Trail
Lilburn, GA 30247

Library of Congress Cataloging-in-Publication Data

Wylie, David, 1953-
    New refrigerants for air conditioning and refrigeration systems /
David Wylie, James W. Davenport.
    p. cm.
    Includes bibliographical references (p.   -   ) and index.
    ISBN 0-88173-224-9
1. Refrigerants. I. Davenport, James W., 1949-   . II. Title.
TP492.7.W95   1996   621.5'64--dc20   96-23417
                                                       CIP

Published by The Fairmont Press, Inc.
700 Indian Trail
Lilburn, GA 30247

Printed in the United States of America

10 9 8 7 6 5 4 3 2 1

ISBN 0-88173-224-9   FP

ISBN 0-13-268715-1   PH

While every effort is made to provide dependable information, the publisher, authors, and
editors cannot be held responsible for any errors or omissions.

Distributed by Prentice Hall PTR
Prentice-Hall, Inc.
A Simon & Schuster Company
Upper Saddle River, NJ 07458

Prentice-Hall International (UK) Limited, London
Prentice-Hall of Australia Pty. Limited, Sydney
Prentice-Hall Canada Inc., Toronto
Prentice-Hall Hispanoamericana, S.A., Mexico
Prentice-Hall of India Private Limited, New Delhi
Prentice-Hall of Japan, Inc., Tokyo
Simon & Schuster Asia Pte. Ltd., Singapore
Editora Prentice-Hall do Brasil, Ltda., Rio de Janeiro

Table of Contents

Tables

Tables (continued)

Figures

Figures (continued)

ACKNOWLEDGEMENTS

With thanks to Doug Scott, VaCom Technologies for assistance with Chapter 4 and overall quality assurance; to James Wylie, Jr. for overall review and valuable chemistry-related information; to Bill Phaklides and Ray Allen at Cal Poly San Luis Obispo, David's professors in the 1960s and 1970s; to Lisa A. McLain, for providing support, graphics, and review; and to Stuart Crumpler for content review.

Introduction

The purpose of this book is to discuss the relevant issues and choices available regarding current and future refrigerants used in air-conditioning and refrigeration equipment.

This book is intended for owners and managers of businesses who are facing choices—dictated by energy use, regulations, and economic factors—as they prepare for the phaseout of certain refrigerants and for the implementation of suitable system options.

Please note that information in this book is based on technical data and tests believed to be reliable, and is derived from sources believed to be reliable. However, the information is subject to change, must be verified before use, and is not warranted by the sources or the authors.

The information contained in this book is not intended as a substitute for proper maintenance, intelligent engineering by persons with technical skill, compliance with regulations, or common sense. Anyone who uses the information in this book does so at their own discretion and risk. Because many factors are outside of the authors' control, we can assume no liability for results obtained or damages incurred through the application of the data presented.

THE SITUATION

As most of us now know, refrigerants, solvents, and other industrial chemicals that contain chlorine, including CFCs (chlorofluorocarbons) and HCFCs (hydrochlorofluorocarbons), are being phased out.

We also know the reasons: when these substances are released into the atmosphere, the chlorine they contain interacts with the Earth's ozone layer and cause its depletion. This depletion reduces the protection from the sun's ultraviolet radiation that the ozone layer provides, and can cause crop failures, skin cancer, and other problems.

The Nobel prize committee has recently acknowledged work in this area by presenting the Nobel chemistry prize to F. Sherwood Rowland of the University of California, Irvine, Mario Molina, now of MIT, and Paul Crutzen of Germany's Max Planck Institute for Chemistry. These are the scientists who first warned of ozone depletion and were presented their prizes for discovering that substances used in refrigerators, air conditioners, and spray cans were threatening Earth's atmosphere.

In addition, there are many direct and indirect costs associated with CFCs. For example, the U.S. Environmental Protection Agency (EPA) has estimated the economic impact of one pound of released CFC to be about $75, considering only the resulting increase in skin cancer.

Also of concern is the global warming effect caused by the release of these substances. There are indications that global warming has caused stratospheric cooling, which makes chlorine even more effective at depleting the ozone layer. Recent reports cite an increase in the levels of greenhouse gases that could drive the U.S. toward mandated limits.[1]

CFC ATMOSPHERIC LEVELS CONTINUE TO INCREASE

CFC levels in the atmosphere are continuing to increase, despite recent actions to limit their production and use. In addition, they appear to disperse globally, as indicated in the *ASHRAE Journal* (American Society of Heating, Refrigerating, and Air-Conditioning Engineers, Inc.) in an article by Sherwood Rowland at the University of California, Irvine. In the ASHRAE article, Rowland claims that recent findings show that concentrations of the refrigerant R-11 are uniform over the entire globe at lower altitudes in spite of the fact that most is released in the Northern hemisphere.[2]

However, there is some encouraging news. The Earth's ozone layer is benefiting from international treaties and mandated CFC phaseout. Recent scientific studies show that the rate of increase of atmospheric CFC levels is lessening.[3] But the *growth* rate will stop only when the release of CFCs stops.

Several predictions indicate that directly because of CFC regulations:

- Ozone levels will return to those measured in 1979 in about 25 years.

- The ozone layer will fully recover (reach its "pre-pollution" levels) by 2050, or in about 55 years.

- Ozone levels will return to normal levels 25 years faster than earlier predicted.

Despite the increasing size of the hole in the ozone layer over the Antarctic continent, we are making progress, and will continue to make progress as long as everyone follows the rules.

The graph below shows the predicted levels of cumulative stratospheric chlorine ("Atmospheric Chlorine Loading") for specific substances in the atmosphere over a 60-year span of time from 1985 to the year 2050.[3] This graph stresses the importance of the phaseout of CFCs and halons, and it demonstrates the smaller role played by HCFCs. Chlorine and bromine levels in the troposphere reached a peak in 1994. It appears at this time that levels of other ozone-depleting substances in the stratosphere will now peak by 1998 and then the recovery begins.

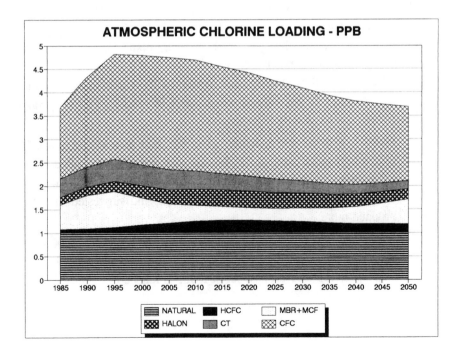

In addition to the natural levels of atmospheric chlorine and CFCs and HCFCs, the levels of halon, carbon tetrachloride (CT), and methyl bromide (MBR) plus methyl chloroform (MCF), other compounds known to destroy ozone, also are shown.

Methyl bromide (used as a solvent and fumigant) and methyl chloroform (used as a cleanser for metal and plastic molds) are always naturally present in the atmosphere. In the graphic on the previous page, the naturally occurring levels of methyl bromide and methyl chloroform in the upper atmosphere occur at and below 1 ppb (parts per billion).

The Antarctic ozone hole appeared at 2 ppb of the chlorine equivalent in the upper atmosphere. One of the goals of the regulations is eventually to reduce the total equivalent chlorine levels to at least 2 ppb or below.

MANY CFC USERS ARE NOT ADDRESSING THE ISSUES

To some extent, most people are aware of the consequences of ozone depletion and global warming as well as the changes that addressing these issues will require. However, a survey reported in the April 1994 issue of the *Air Conditioning, Heating & Refrigeration News* shows that a total of 58% of the customers of HVAC/R industry contractors are "not very" or "not at all prepared" for the upcoming phaseout of CFCs. That is, their customers have no plans for replacing, converting or retrofitting existing equipment.[4]

CFC users will eventually need to address the situation because current and anticipated regulations will cause CFC and HCFC refrigerants to become scarcer and more expensive. For example:

- DuPont, a major manufacturer of CFCs, had planned to stop CFC production at the end of 1994—one year earlier than called for in the regulations. However, they in fact did manufacture more CFCs because the EPA reevaluated the situation, and suggested that they produce more.

- Estimates by chiller manufacturers indicate between 50,000 and 60,000 units in the U.S. will be using CFC refrigerants when the January 1, 1996 CFC phaseout deadline is reached.

- Shortages of refrigerants are predicted along with "a mad scramble to do something." And, as a result of the production phaseout, CFC costs might increase dramatically.

What is being done at your facility? Are you prepared for the upcoming shortage and increasing cost of CFC and HCFC refrigerants?

ISSUES CFC USERS MUST FACE

As the individuals responsible for the use of refrigerants in your buildings or businesses, you and your associates must deal with these issues and face decisions that are often confusing and complex. Now is the time to answer some very important questions about the future:

- Have you taken any action toward developing a plan for managing refrigerants?
- Will you be caught short on your supply of refrigerants? Should you begin or continue to stockpile refrigerant supplies?
- Will you find that manufacturers of alternative equipment and equipment repair services have such a backlog that they are unable to meet your needs?
- Are you informed about the fundamental issues and options, including your legal responsibilities and liabilities?
- Are you overwhelmed with conflicting or contradictory information?

The CFC issues affect the air-conditioning and refrigeration sectors of the industry differently. HVAC systems, whether small package units or large centrifugal chillers, are usually purchased from catalogs as pre-engineered, off-the-shelf items from one of several manufacturers, and the mechanical technology and refrigerants used are predetermined.

In contrast, most commercial and industrial refrigeration systems are custom-designed—often by companies whose prime focus may be display fixtures or processing equipment. As a result, pre-engineered solutions are less likely to exist for refrigeration systems.

This means users of refrigeration equipment, such as grocery stores and cold storage facilities, face a more complex set of choices in their search for individual, custom-designed and custom-built solutions. This sector needs to consider not only the refrigerant but a wide variety of machinery options.

When preparing for the future, you must plan carefully. No single solution will suit all situations.

What are the risks of waiting? If you procrastinate, you may face a situation that is expensive and uncomfortable. On the other hand, don't rush to implement the least-expensive solutions to meet minimum CFC compliance that may potentially increase energy consumption and demand. Seek a practical balance between complying with the regulations, doing the right thing by the environment, and making economic decisions that are viable for your company.

A PREVIEW OF THE BOOK'S TOPICS

This book will present information we hope will help you as you and your associates make important decisions for meeting future refrigeration and air-conditioning needs. This book contains the following chapters and topics:

- **Chapter 1—Regulatory Issues:** This chapter provides an overview of regulatory international treaties, and the current and future national and local regulations to which you must respond.

- **Chapter 2—Refrigerant Options:** Part of the decision you must make relates directly to the alternative refrigerant options available. This chapter discusses issues relevant to existing and future alternative refrigerants, their properties, applications, supply, demand, and costs.

- **Chapter 3—Machinery Choices For Air-Conditioning Systems:** There are many different machinery choices available for new and upgraded air-conditioning systems. In this chapter we look at chiller systems and the impact new refrigerants have on the types of compressors they use, their operating pressures, system efficiency, and other factors. This chapter also includes a sample system evaluation, findings, and proposed design, and a list of chiller system manufacturers.

- **Chapter 4—Machinery Choices For Refrigeration Systems:** This chapter discusses the many machinery and refrigerant choices available for commercial refrigeration systems for conversions and new equipment. Also, this chapter provides research criteria and results of a refrigeration demonstration project, the description of a "prototype" refrigeration system, and the results of tests performed in an operating store conducted by the Electric Power Research Institute (EPRI).

- **Chapter 5—Developing a Refrigerant Management Plan:** This chapter provides suggestions on the key issues you must consider as you implement a plan for the phaseout of CFCs. Also in this chapter, we will examine what some companies have done regarding establishing a corporate policy and producing a refrigerant management plan.

- **Glossary**—The Glossary provides definitions of common terms, abbreviations, and acronyms.

Chapter 1

Regulatory Issues

INTRODUCTION

All refrigerants are regulated to some extent. There are international, national, state, and local regulations for many aspects of how refrigerants are used, stored, handled, and replaced. The overabundance of regulatory bodies and new laws regulating the HVAC/R industry can be confusing, but cannot be ignored.

It is essential that you are aware of the regulations that pertain to your business or industry—there are serious legal and economic ramifications if you don't comply or are not prepared to comply. Regulatory bodies are stepping up their efforts and are giving out heavy fines for noncompliance. Also, there are potential lawsuits from anyone who can prove that lack of proper refrigerant handling has posed a health or safety hazard. To make things even more complicated, sometimes just knowing the regulations is often not enough; we'll see how what's known as "industry consensus practice" can supersede local codes.

In this chapter, we will discuss the reasoning behind the regulations and the issues you must be aware of and consider in your planning. The regulations we will discuss in this chapter cover the most common refrigerants: CFCs (chlorofluorocarbons), HCFCs (hydrochlorofluorocarbons), and HFCs (hydrofluorocarbons) which do not contain chlorine.

THE BASIS FOR THE REGULATIONS

The regulations governing refrigerants are based on the potentially harmful properties and effects of refrigerants, including long-term environmental impact (such as ozone depletion and the "greenhouse effect" or global warming) as well as immediate health and safety issues such as toxicity and flammability.

Ozone Depletion

In the Earth's upper atmosphere (specifically, the stratosphere and troposphere), energy from ultraviolet light turns atmospheric oxygen into ozone. Ozone is oxygen (O_2) with three atoms in each molecule (O_3) instead of the usual two atoms. Upper atmosphere ozone provides the Earth with a layer of protection from harmful ultraviolet radiation from the sun.

Scientific research indicates that CFCs destroy ozone when the chlorine they contain interacts with ozone. One CFC molecule can destroy 100,000 molecules of ozone; ten chlorine atoms can destroy a million ozone molecules.

A chain of reactions begins when ultraviolet light strikes a CFC molecule in the stratosphere, freeing a chlorine atom. The free chlorine atom combines with other molecules to form new compounds. In particular, the free chlorine atom attacks an ozone molecule breaking it apart by taking one of the ozone's three oxygen atoms. This creates a molecule of ordinary oxygen (with two atoms) and a molecule of chlorine monoxide.

A free oxygen atom, that would usually join with an oxygen molecule to form ozone, then reacts with and breaks up the chlorine monoxide molecule. In the process, another oxygen molecule is formed and the chlorine atom is released. As this series of events occurs, the chlorine both destroys an ozone molecule and prevents a new one from forming; and it is free to repeat the process.

Evidence Relating CFCs and HCFCs to Ozone Depletion

For most reasonable people, ozone-depletion science is well established and accepted, and it has been confirmed by observed data. Despite the arguments of skeptics, CFCs have reached the upper atmosphere in sufficient amounts to cause ozone depletion. Ozone depletion has been confirmed to be linked to man-made emissions, and it also is linked to changes in the climate.

NASA satellite measurements show that the levels of hydrogen fluoride found in the atmosphere directly correspond to CFC levels; they are caused by man and are not from natural occurrences such as volcanoes. Also, only one-sixth of the detected levels of chlorine are from natural sources; the rest are from man-made chemicals such as CFCs.[5]

Other Ideas and Opinions

Contrary ideas and opinions exist. There are some who disagree with the scientific findings and some who would offer perhaps unusual approaches to solving the ozone-depletion problem.

The governor of Arizona Fife Symington recently said that the bans on Freon production are based on "hokey science." He went so far as to sign a bill that would legalize the manufacture, sale, and use of Freon after December 31, 1995.[6]

The most prevalent approaches to reducing ozone depletion are the international treaties, which drive the national and local regulations governing the production and use of chemical compounds that have the potential to deplete ozone. There also are other rather curious (perhaps odd-ball) approaches to the problem.

For example, UCLA Physics Professor Alfred Wong feels that one way to decrease the depletion of ozone is to artificially provide the extra electrons that chlorine demands.

In a laboratory setting, Professor Wong has simulated conditions found in the Earth's stratosphere and troposphere and has experimented with adding electrons to the chlorine atoms in CFCs. He has demonstrated that when provided with extra electrons, chlorine atoms become inert and no longer consume ozone.

Professor Wong envisions a method of creating and emitting electric charges into the atmosphere that involves using the sun's energy in conjunction with a metal sheet or "electrical curtain" supported by navigable airship balloons. This platform would emit electric charges into the air and the chlorine atoms would acquire the extra electrons they seek from this reaction instead of from the ozone.[7]

Well, if professor Wong or anyone else can make such a project scientifically feasible and cost effective (if the costs for the reduction of ozone-depleting chlorine are less than $75 per pound), then we say, "let it fly!"

While it is important to explore alternative solutions to this "man-made" problem, the prevailing, cost-effective solutions are those that investigate the use of alternative refrigerants and containment methods.

Measuring Chemicals' Ozone Depletion Potential (ODP)

Donald Wuebbles, of the University of Illinois, Urbana, originated the ozone depletion potential (ODP) concept.

ODP is defined as the relative ability of a substance to destroy stratospheric ozone and is determined relative to CFC-11 which has an ODP value of one. The higher the number the greater the substance's ODP.

Substances are divided into two classes based on their ODP by the U. S. Clean Air Act (as specified in the U. S. Regulatory Update January 1994):

- **Class I** compounds *significantly* cause or contribute to harming the ozone layer. The U.S. Environmental Protection Agency (EPA) is required to add compounds with an ODP greater than or equal to 0.2 to the list of Class I substances.

  These substances, which include all isomers, are further classified into seven groups. The only group of these seven that we are concerned with is Group 1. Group 1 includes the refrigerants CFC-11, CFC-12, CFC-113, CFC-114, and CFC-115 (used to make R-502).

- **Class II** compounds are those that are known, or may be reasonably anticipated, to cause or contribute to harmful effects on the ozone layer, but have an ODP less than 0.2. Class II substances currently include only HCFCs and their isomers that have one, two, or three carbon atoms. (A brief chemistry lesson is presented in Chapter 2).

The following table lists Class I regulated substances and gives their ODP rating.[8] (In Chapter 2, the ODP ratings of newer refrigerants are provided.)

Ozone Depletion Potential (ODP) of Class I Substances			
Substance	ODP	Substance	ODP
CFC-11	1.0	CFC-13	1.0
CFC-12	1.0	CFC-111	1.0
CFC-113	0.8	CFC-112	1.0
CFC-114	1.0	CFC-211	1.0
CFC-115	0.6	CFC-212	1.0
Halon 1211	3.0	CFC-213	1.0
Halon 1301	10.0	CFC-214	1.0
Halon 2402	6.0	CFC-215	1.0
Carbon tetrachloride	1.1	CFC-216	1.0
Methyl chloroform (1,1,1-trichloroethane)	0.1	CFC-217	1.0

Table 1-1   Ozone Depletion Potential (ODP) of Class I Substances

Global Warming and the Greenhouse Effect

The greenhouse effect likens the Earth's atmosphere to the inside of a greenhouse. The glass windows of a greenhouse let in sunlight and the sunlight warms up the objects inside the greenhouse which then radiate heat. All of the heat is captured inside and the temperature rises. The Earth and its atmosphere are like a greenhouse; and, in the same way as the glass of a greenhouse, the Earth's atmosphere lets in radiation (short-wave and visible solar radiation).

Some of the radiation energy bounces off the Earth and is radiated up to the atmosphere. Some of this radiation escapes the atmosphere and returns to outer space. Other radiation is absorbed by carbon dioxide (and other so-called "greenhouse gases") and water vapor contained in the atmosphere causing the Earth and its atmosphere to warm up.

CFCs, in addition to destroying stratospheric ozone, have been recognized as contributing to increases in global warming. Other greenhouse gases such as methane, oxides of nitrogen, and hydrofluorocarbons (HFCs) also contribute to this effect.

There is a direct relationship between the amount of greenhouse gases that are in the atmosphere to how much heat is retained. Accumulations of these gases affect the "radiative balance" of the Earth-atmosphere system and affect, to some degree, the global climate.

The Controversy Regarding Global Warming and the Greenhouse Effect

Scientists are unsure exactly where greenhouse gases originate, and where and how they collect in the atmosphere. Also unclear is exactly how these gases interact with clouds, oceans, polar ice, and other physical features of the Earth. That is, they are uncertain of the degree to which the global climate is affected by water vapor, carbon dioxide, and greenhouse gas emissions. Also, there is uncertainty about what is the best way to interpret climate data since there is a complex mix of factors involved. Scientists can't even agree whether the atmosphere is warming or cooling or how fast the average temperature is changing.

One point of view predicts significant, long-term climatic changes on Earth, with dramatic changes in global climate patterns as early as the year 2000. If the quantity of greenhouse gases continues to rise, before the year 2100 global average temperatures may rise as much as 9° F (5° C). This rise in temperature would alter weather patterns, cause extremes of drought and rainfall, and would harmfully interrupt food production. It would also cause the polar ice caps to melt, resulting in substantially higher water levels along the shorelines of oceans and flooding in coastal areas. Scientists point to a recent crack in the Ross Ice Shelf in the Antarctic and to a 1.5° F increase in the average temperatures in the Alps. Whether these are natural phenomena or directly result from global warming effects is debated.

On the other hand, other scientists think that other factors, such as ash from volcanic eruptions and man-made dust in the atmosphere, might **reduce** the overall effect of increases in greenhouse gas levels by increasing the reflection of the sun's radiation. This may prevent temperatures from rising or may tend to lower the temperature on Earth.

Also, when comparing the relative influences of specific factors on global warming and the greenhouse effect, scientists disagree about whether water vapor or carbon dioxide plays the more significant role. Of the total global carbon dioxide emissions, only about 5% is from man-made sources. And of that 5%, only a small fraction results from burning fossil fuels, wood, and other fuels directly generated from plants.[9]

Measuring Chemicals' Potential Effect on Global Warming (GWP and TEWI)

There are two common measures of a compound's potential impact on global warming: Global Warming Potential and Total Equivalent Warming Impact.

Global Warming Potential (GWP)

The direct effect of CFCs or other substances on global warming is measured as the Global Warming Potential (GWP).

The Global Warming Potential indicates the relative ability of a greenhouse gas to trap heat (or absorb energy) together with the amount of the gas present in the atmosphere over a specified period of time.

GWP is an index that provides a simplified way of describing the relative ability of each man-made greenhouse gas emission to affect future "radiative forcing." Radiative forcing relates to the direct heat-trapping effect of greenhouse gases. It is defined as factors that can disturb the radiative balance of the Earth-atmosphere system and thereby affect global climate change by causing warming, or in the case of negative radiative forcing, global cooling.

Another important aspect is that GWP is directly related to the "atmospheric lifetime" of a substance. That is, how long it can survive in the atmosphere before decomposing or reacting with other compounds. GWPs for different gases are given over different Integration Time Horizons, typically 20 years, 100 years, and 500 years. Be aware that although the Integration Time Horizon is an important factor regarding the effect of direct emissions of greenhouse gases, GWP values will differ for different substances over different time horizons.

To present one example, the GWP of CFC-12 is 7100 in a 20-year Integration Time Horizon. This means one pound of CFC-12 released all at once into the atmosphere is equivalent to the radiative forcing of 7100 pounds of carbon dioxide, during the first 20 years after its release. Using a 500-year time horizon, the GWP of CFC-12 is 4500.[10]

GWP is typically measured relative to either CFC-11 or CFC-12 or relative to carbon dioxide. GWP values based on carbon dioxide have been determined by the Intergovernmental Panel on Climate Change

Global Warming Potential (GWP) of Gases			
	Integration Time Horizon		
Compound	20 yrs.	100 yrs.	500 yrs.
Carbon dioxide	1	1	1
Methane	63	21	9
Nitrous oxide	270	290	190
CFC-11	5000	4000	1400
CFC-12	7900	8500	4200
CFC-113	4500	4200	2100
CFC-114	6000	6900	5500
CFC-115	6200	9300	13000
HCFC-22	4300	1700	520
HCFC-123	300	93	29
HCFC-124	1500	480	150
HCFC-141b	1800	630	200
HCFC-142b	4200	2000	630
HFC-32	1800	580	180
HFC-125	4800	3200	1100
HFC-134a	3300	1300	420
HFC-143a	5200	4400	1600
HFC-152a	460	140	44

(IPCC). When measured relative to CFC-11 or CFC-12, it is called Halocarbon GWP (HGWP).

The table above indicates several common compounds' GWP (global warming potential relative to carbon dioxide) in terms of their Integration Time Horizons.[11]

Total Equivalent Warming Impact (TEWI)

Total Equivalent Warming Impact (TEWI) is defined as the sum of the direct (chemical action caused by refrigeration emission) and indirect (energy-related) emissions of greenhouse gases, from the use of

CFC or fluorocarbon refrigerants. TEWI is the most accurate overall measure of global warming since it considers the total global warming impact of the whole refrigeration system, including the energy it uses, and not just the GWP of a particular refrigerant.

The direct emissions of refrigerants typically account for a small part of the TEWI factor. There is sizable indirect global warming from the carbon dioxide created by the combustion of fossil fuel in generating the electrical power to run refrigeration equipment.

TEWI shows we must consider both the direct and indirect emissions caused by air-conditioning and refrigeration systems, and emphasizes the importance of energy efficiency and minimizing leaks. The TEWI calculations consider such things as system refrigerant charge, annual leakage rate, annual operating hours, refrigerant GWP, and annual energy consumption. For refrigeration and air-conditioning equipment, TEWI is computed by estimating the amounts of refrigerant released to the atmosphere, the annual energy use of the equipment, and the typical useful lifetime of the equipment.

This means a refrigerant with a higher GWP can generate a lower TEWI if the equipment used in a refrigeration or air-conditioning system is more efficient and measures have been taken to prevent refrigeration leaks. In these cases, from 96% to 99% of the TEWI is from the carbon dioxide from energy consumption (power plant output). So if TEWI is to be reduced further, power companies need to make their power plants more efficient, and refrigeration and air-conditioning manufacturers need to make their equipment more efficient.[12]

A Study of Global Warming

A study was performed by scientists and others at the Oak Ridge National Laboratory in conjunction with Arthur D. Little, Inc. and sponsored by the Alternative Fluorocarbons Environmental Acceptability Study (AFEAS) and the U.S. Department of Energy (DOE). The study presented information on the impact of alternate refrigerant technologies (those that will be used to replace CFCs) on energy use and global warming.[13]

The study attempted to answer exactly how one can measure the relative contributions of greenhouse gases to direct heat trapping effects. The study considered GWP and TEWI among many other indices.

How much a greenhouse gas contributes to the calculated global warming depends on several factors:

- The amount of the gas that is emitted into the atmosphere.
- The length of time that elapses before the gas is removed from the atmosphere.
- The energy absorption properties of the gas.

The "applications" analyzed were carbon dioxide, methane, nitrogen oxides, and many different common refrigerants including CFCs, HCFCs, and HFCs.

The authors of the study considered GWPs relative to carbon dioxide. To determine TEWI values, their calculations of the amount of gases emitted by these "applications" used estimates of either direct fluorocarbon emissions, or the amount of carbon dioxide emitted as a result of the energy used.

The calculations also used the parameters of GWP values. In this way, refrigerant emissions were "converted to their equivalent emissions of carbon dioxide so that there is a common basis for comparing impacts."

Because carbon dioxide, CFCs, HCFCs and HFCs are eliminated from the atmosphere at very different rates, they considered past concentration changes in the atmosphere in determining the relative contributions of the various greenhouse gases.

As we saw earlier, the specified period of time used to calculate GWP values is the Integration Time Horizon. Because Integration Time Horizons of 20 years and 100 years don't present a complete picture regarding the potential effects of both short-lived substances (HCFCs for example) and longer-lived substances (such as carbon dioxide), the authors of the "Energy and Global Warming Impacts of CFC Alternative Technologies," report used GWPs calculated at the 500-year Integration Time Horizon with other calculations at the 100-year Integration Time Horizon.

Major Findings of the Study

The bar graph in Figure 1-1 gives the results of the study's analysis for the 100-year TEWI. The figures show two kinds of data: the changes in TEWI of the CFC alternatives to the baseline CFC technology, and the relative proportions of the effects of direct and indirect sources that make up the TEWI value.

The major findings of the study addressed all of the major use categories, including refrigerants, insulation, and cleaning substances:

- Replacing CFCs with HCFCs or HFCs reduces TEWI for all CFC end-use applications considered in the report. Depending on the application, TEWI values can be reduced from 10% to 98%. Improvement in TEWI is especially noticeable in the applications where emissions are high, such as solvent cleaning, automotive air conditioning, and commercial building roof insulation.

- Regarding the HCFC and HFC alternatives, the relative contributions of indirect and direct emissions to TEWI values varies substantially. For the applications that are energy efficient (low indirect source emissions) but leak refrigerant (high direct source emissions) and therefore have a high TEWI, replacing these systems with equipment that does not leak or repairing the current equipment so it doesn't leak can reduce the TEWI dramatically. This applies especially to retail refrigeration equipment. In short, equipment must be designed and constructed better and manufactured with better materials.

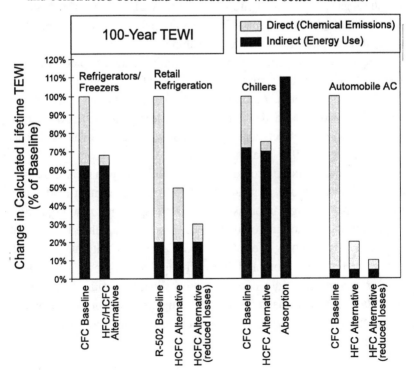

Figure 1-1 100-Year Total Equivalent Warming Impact (TEWI)

Refrigerants and the Global Warming Issue: Some Conclusions

There are those who have enough evidence to be convinced that refrigerants are a contributing factor to global warming. And, there are those who need more data or evidence to be convinced. This latter camp argues that government policies dictating actions based on global warming have already overtaken the science and should be reduced to a smaller scale, over a shorter time frame. This, they argue, will prevent what may be unnecessary overreactions.

Time will tell the true impact an industrial society has had on global warming, but there is general acceptance that greenhouse gases can trap infrared radiation and prevent it from being released into space and that the presence of these gases upsets the radiative balance of the Earth-atmosphere system. Most agree that the atmosphere is warmer with the addition of these gases than it would be without them.

But because there are no hard and fast conclusions that can be drawn, and because what we know about global warming is outweighed by what we don't know, we suggest a prudent path.

The legal provisions regarding CFCs were also ahead of the science and some thought (and some still think) the policy was an overreaction. But as it turned out, it was the correct thing to do.

The actual documented data demonstrating ozone depletion arrived after the fact, about the same time the effects of the regulations began to realize positive benefits. As a result, NASA estimates that if the regulators had waited for the data, it may have taken another 100 years to realize the positive effects already obtained: that ozone levels will reach those measured in 1979 in about 25 years.[5]

GWP parameters do provide scientists and policy makers a measure of possible future global warming impacts from current man-made emissions. But before large amounts of money are spent and new regulations are passed, more scientific evidence needs to be accumulated and examined. Further restraining the interim HCFC and HFC alternatives might be counterproductive from a global warming point of view.

Our recommendation: common sense should prevail. Where possible, only low-cost measures should be implemented such as a ban on intentional venting of HFCs. We would expect the EPA to implement such a ban.

At the same time the air-conditioning and commercial refrigeration industry can be proactive in their efforts to maximize system efficiency, to improve maintenance, and to promote ways to minimize refrigerant leaks.

The Ozone Depletion Potential and Global Warming Potential of Common Refrigerants

The graphic below illustrates the ODP and GWP of conventional and alternative refrigerants.

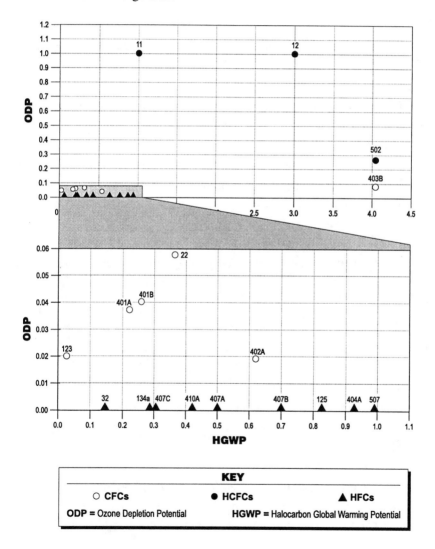

Figure 1-2 ODP and HGWP for Conventional and Alternative Refrigerants

Safety Issues

The two factors that are of key concern relative to refrigerant safety are the refrigerant's toxicity and flammability. In addition, high-level exposure to refrigerant—such as through serious leaks or spills—can cause heart attacks or cause unconsciousness and asphyxiation.

These issues are considerations for both traditional and new substitute refrigerants.

Terminology of Toxicity and Flammability

Because many of the standards refer to a compound's toxicity and flammability levels, it is useful to have a basic understanding of some of the terms that are commonly used when referring to these issues.

Toxicity

The EPA reports that the two types of toxicity—chronic and acute—pose completely different kinds of safety hazards, and that all refrigerants pose both types of toxicity.

- **Chronic toxicity** relates to long-term exposures over a lifetime of experience with refrigeration equipment.

- **Acute toxicity** describes the dangers posed by short-term exposure to very high concentrations of refrigerant (as in cases of catastrophic releases) or during very brief but intense exposures that service personnel may accidentally experience during repairs (such as opening a compressor).

The Programme for Alternative Fluorocarbon Toxicity Testing (PAFT) is a cooperative effort sponsored by the major CFC producers from many countries. Their task is to expedite the toxicity testing of alternative refrigerants, which could take many years to complete.

Measures of Chronic Toxicity

To assess the hazard of chronic exposure, it is necessary to consider how much of a substance a person can be exposed to over time without ill effect. This is typically expressed as a the Threshold Limit Value for a Time-Weighted Average (TLV-TWA).

- The **Threshold Limit Value**® (**TLV**) refers to how much of a substance (usually expressed in ppm (parts per million) airborne concentration of contaminants) most workers may be exposed to during

---

®TLV is a registered trademark of the American Conference of Governmental Industrial Hygienists.

a normal 40-hour work week throughout a normal working career—without suffering adverse effects.

- The **Time-Weighted Average (TWA)** refers to the average concentration of a substance over a period of time. It takes into account the fact that the actual, specific amount of airborne concentration in an area will vary throughout the day.

The higher the TLV-TWA rating of a compound, the less toxic that compound is because the TLV-TWA rating indicates the concentrations you can be exposed to safely.

Other indicators of chronic toxicity include:

- The **Acceptable Exposure Limit (AEL)**—a value set by refrigerant manufacturers indicating the concentration a worker could be exposed to eight hours a day for a working lifetime without ill effects. (The AEL is a preliminary rating used until a value is given by the American Conference of Governmental Industrial Hygienists.)

- The **Permissible Exposure Level (PEL)**—A value set by OSHA (the U.S. Occupational Safety and Health Administration) indicating the maximum TWA concentration of a substance allowed during any eight-hour work shift.

Measures of Acute Toxicity

The head toxicologist of the EPA reports that "...it is acute toxicity, not chronic exposure, which represents the greater danger to the HVAC industry's operators and service people."

The most common rating of acute toxicity is the **Immediately Dangerous to Life and Health (IDLH)** rating. IDLH values are set by the U.S. National Institute of Occupational Safety and Health (NIOSH). They indicate the highest concentrations most adults can be exposed to and still be able to escape within 30 minutes without using a respirator and without suffering from an impaired ability to escape (such as unconsciousness or severe eye irritation) or from irreversible effects on their health.

Another rating of acute toxicity is the **Short-Term Exposure Limit (STEL)**. STEL is a rating that represents the highest concentrations most adults can be exposed to and still be able to escape within a matter of minutes.

Flammability

Two refrigerant flammability ratings are the **Lower Flammability Limit (LFL)** and the **Heat of Combustion** established by the American Society of Heating, Refrigerating, and Air-Conditioning Engineers (ASHRAE). These ratings are determined using specific laboratory tests performed under specific testing conditions to provide relative flammability values (as dictated by ASHRAE Standard 34-1994).

Toxicity Levels for Various Refrigerants

Refrigerant manufacturers typically provide Material Safety Data Sheets (MSDSs) that indicate the toxicity level of a refrigerant, outline the risks associated with the chemical, and provide safety guidelines and first-aid procedures. These MSDSs are updated frequently to reflect the current ratings and recommended safety measures associated with a refrigerant.

An example MSDS (courtesy of ICI) is reproduced on the following pages. Anyone who works with or around refrigerants should be familiar with the information on the MSDS for a refrigerant. In addition, current recordkeeping documents, required for most chemicals used on site, should be kept readily available for reference.

Material Safety Data Sheet

---

1. Chemical Product and Company Identification   3640

ICI Klea

Concord Plaza, 3411 Silverside Road   Issue Date: 03/16/95
P.O. Box 15391   Rev.: G
Wilmington, DE 19850   BPCS: 274

ICI Operator (24 hr.): (302) 887-3000

Medical Emergency (24 hr.): (800) 228-5635 Extension 181***

Chemical Emergency (24 hr.) Involving Transportation
Spills, Leaks, Fires or Accidents: (800) 424-9300

KLEA® 134a

General Use: Refrigerant
Alternative names: Fluorocarbon 134a, R134a, HFC 134a, HFA 134a

---

2. Composition/Information on Ingredients

Ingredients	%	OSHA PEL
1,1,1,2-Tetrafluoroethane (CAS 811-97-2)	100	Not Listed

Ingredients not precisely identified are proprietary or nonhazardous. Values are not product specifications.

---

3. Hazards Identification

Emergency Overview:
  Appearance: Colorless liquefied gas with faint ethereal odor
  Physical hazards * : Compressed liquefied gas
  Health hazards * : Harmful (central nervous system depression, cardiac arrhythmias)
  * Hazard summary as defined by OSHA Hazard Comm. Std., 29 CFR 1910.1200.

Potential Health Effects:

  General: The health hazard assessment is based on toxicity studies together with information from a search of the scientific literature and other commercial sources.

  Ingestion: Extremely unlikely to occur in use.

  ---continued---

Figure 1-3 Sample Material Safety Data Sheet

Hazards Identification (continued):

Eye contact: Liquid splashes or vapor spray may cause freeze burns.

Skin contact: The liquid form of this product may cause freeze burns (frostbite-like lesions).

Skin absorption: This product will probably not be absorbed through human skin.

Inhalation: Exposure to very high vapor concentrations can induce anesthetic effects progressing from dizziness, weakness, nausea, to unconsciousness. It can act as an asphyxiant by limiting available oxygen. At very high doses, cardiac sensitization to circulating epinephrine-like compounds can result in fatal cardiac arrhythmias.

Other effects of overexposure: None expected.

4. First Aid Measures

Skin: Thaw affected area with water. Remove contaminated clothing. Caution: clothing may adhere to the skin in the case of freeze burns. After contact with skin, wash immediately with plenty of warm water. If symptoms (irritation or blistering) develop, get medical attention.

Eyes: Immediately flush with plenty of water. After initial flushing, remove any contact lenses and continue flushing for at least 15 minutes. Have eyes examined and treated by medical personnel.

Ingestion: Not applicable.

Inhalation: Remove victim to fresh air. Keep warm and at rest. If not breathing, give artificial respiration, preferably mouth-to-mouth. If breathing is labored, give oxygen. In the event of cardiac arrest, apply external cardiac massage. Do not administer adrenaline or similar sympathomimetic drugs as cardiac arrhythmias may result. Get immediate medical attention.

5. Fire Fighting Measures

Flashpoint and method: Does not flash

Autoignition temperature: Not applicable

Flammable limits (STP): Nonflammable

General hazards: Compressed liquefied gas. HFC 134a is not flammable in air under ambient conditions of temperature and pressure. In laboratory tests, under conditions of high pressure, HFC 134a/air mixtures were shown to be flammable. In general, for the test equipment used, at temperatures up to 170 deg. C, flammable mixtures were only produced at pressures greater than 50 psia, and with more than 50 volume % air. Mixtures of HFC 134a should not be used for pressure or leak testing. Thermal decomposition will evolve toxic and irritant vapors.

---continued---

Figure 1-3 Sample Material Safety Data Sheet (continued)

Fire Fighting Measures (continued)

Firefighting Instructions: *Not applicable. Use media suitable for surrounding fire. Use water spray to cool containers.*

Firefighting equipment: *Self-contained breathing apparatus with full facepiece and protective clothing.*

Hazardous combustion products: *Heavy vapors can suffocate. Highly toxic decomposition products.*

6. Accidental Release Measures

Shut off leak if without risk. Ventilate the spill area. If possible dike and contain spillage. Prevent liquid from entering sewers, sumps or pit areas, since vapor can create suffocating atmosphere. Use self-contained breathing apparatus to avoid suffocation. Allow spilled liquid to evaporate. Protect against frost-bite from evaporating liquid.

7. Handling and Storage

Storage temperature: *Keep at temperature not exceeding 113 deg. F (45 deg. C).*

General: *Keep in a cool place. Keep containers dry. Keep away from direct sunlight, heat and sources of ignition.*

8. Exposure Controls/Personal Protection

Exposure guidelines: *No ACGIH TLV or OSHA PEL assigned. Minimize exposure in accordance with good hygiene practice. ICI has established an employee exposure standard of 1,000 ppm (8hr TWA) for this material.*

Engineering controls: *Ventilate low-lying areas such as sumps or pits where dense vapors may collect. Use ventilation adequate to maintain safe levels. Provide eyewash station in work area.*

Respiratory protection: *Not normally needed, if controls are adequate. If needed, use MSHA-NIOSH approved respirator for organic vapors. For high concentrations and oxygen-deficient atmospheres, use positive pressure air-supplied respirator.*

Protective clothing: *Impervious gloves if any possibility of skin contact with the liquid. Additional protection may be required such as apron, arm covers, or full body suit, depending on conditions.*

Eye protection: *Chemical tight goggles; full faceshield in addition if splashing is possible.*

Figure 1-3   Sample Material Safety Data Sheet (continued)

9. Physical and Chemical Properties

Appearance: Colorless liquefied gas
Boiling point: -15.1 deg. F, -26.2 deg. C
Vapor pressure (mmHg at 20 deg. C): 4268
Vapor density (air = 1): 3.3
Solubility in water: Very low
pH: Not applicable
Specific gravity: 1.27 at 20 deg C
% Volatile by volume: 100

10. Stability and Reactivity

Stability: Stable under normal conditions.

Incompatibility: Finely divided metals, magnesium and alloys containing more than 2% magnesium. Can react violently if in contact with alkali or alkali earth metals such as sodium, potassium or barium.

Hazardous decomposition products: Halogen acids by thermal decomposition and hydrolysis.

Hazardous polymerization: Will not occur.

11. Toxicological Information

Possible Human Health Effects:

Inhalation; High atmospheric concentrations may lead to anaesthetic effects, including loss of consciousness. Very high exposures may cause an abnormal heart rhythm and prove suddenly fatal. Higher concentrations may cause asphyxiation due to reduced oxygen content of the atmosphere.

Skin contact: Liquid splashes or spray may cause freeze burns. Unlikely to be hazardous by skin absorption.

Eye contact: Liquid splashes or spray may cause freeze burns.

Ingestion: Highly unlikely-but should this occur, freeze burns will result.

Animal data:

The inhalation 4 hour LC50 in rats was greater than 500,000ppm HFC134a.

Because of its volatility, this compound has not been tested for skin or eye irritancy, or skin sensitization.

---continued---

Figure 1-3 Sample Material Safety Data Sheet (continued)

Toxicological Information (continued)

The threshold for cardiac sensitization (arrhythmia) in dogs pretreated with epinephrine was an atmosphere of 75,000ppm.

No effect of any kind was seen in a 90-day inhalation study in the rat at dose levels up to, and including, 50,000ppm (6 hours per day, 5 days per week).

No developmental effects were seen in the rabbit following inhalation exposure to 40,000ppm during gestation despite slight maternal toxicity. In a range-finding study in the rabbit, possible minimal embryolethality was seen at a dose level of 50,000ppm. In the rat, slight fetotoxicity was present at an inhalation dose of 50,000ppm administered during gestation and no effects were seen at 10,000ppm. In another study in the rat, no developmental effects were seen at a dose of 100,000ppm in the presence of slight maternal toxicity; clear maternal effects were followed by embryotoxicity and fetotoxicity at a dose level of 300,000ppm. There were no increases in the incidence of fetal malformations in rats or rabbits at doses up to and including 300,000 and 50,000ppm, respectively.

HFC134a showed no genetic toxicity in a range of in-vitro and in-vivo tests.

No adverse effects were found in a study in which rats were followed to week 104 after receiving 300mg/kg bodyweight/day of HFC134a by gavage for 52 weeks. In a 2-year inhalation study in rats, no adverse effects of any kind were observed except increased incidences of non-life threatening, benign, microscopic testicular interstitial (Leydig) cell tumors and associated interstitial cell hyperplasia which were confined to the top dose of 50,000ppm.

12. Ecological Information

Persistence and degradation: Decomposes comparatively rapidly in the lower atmosphere (troposphere). Atmospheric lifetime is 15.6 years. Products of decomposition will be highly dispersed and hence will have a very low concentration. Does not influence photochemical smog (ie. it is not a VOC under the terms of the UNEC convention). Has no effect on the ozone layer.

Effect on effluent treatment: Discharges of the product will enter the atmosphere and will not result in long term aqueous contamination.

13. Disposal Considerations

Disposal method: Discarded product is not a hazardous waste under RCRA, 40 CFR 261.

Container disposal: For disposable (DOT 39) cylinders only: Do not distribute, make available, furnish or reuse empty container when once emptied of the original product. Open valve to remove pressure in the cylinder. Then puncture, drill, crush or otherwise destroy empty cylinder and dispose of in a facility permitted for nonhazardous waste.

Figure 1-3 Sample Material Safety Data Sheet (continued)

(continued)KLEA 134a *Page 6*

14. Transport Information

DOT Hazard Description per HM-181

Proper Shipping Name: 1,1,1,2-TETRAFLUOROETHANE (R134A)
Hazard Class: CLASS 2.2
Identification Number: UN 3159
Packaging Group: None
Hazardous Substance (RQ): None
Placard/Label: NON-FLAMMABLE GAS

15. Regulatory Information

TSCA (Toxic Substances Control Act) Regulations, 40 CFR 710: All ingredients are on the TSCA Chemical Substance Inventory.

CERCLA and SARA Regulations (40 CFR 355, 370, and 372): This product does not contain any chemicals subject to the reporting requirements of SARA Section 313.

The information herein is given in good faith,
but no warranty, expressed or implied, is made.

Prepared/Reviewed: 01/19/93

Figure 1-3 Sample Material Safety Data Sheet (continued)

Chiller Refrigerant Safety Indicators							
	R-11	R-123	R-12	R-134a	R-22	R-407C	R-717
Acute Toxicity (ppm)	1,000	1,000	50,000	75,000	50,000	1000	35
Chronic Toxicity (ppm)	1000 PEL	30 AEL	1,000 PEL	1,000 PEL	1,000 PEL	1000 AEL	50 PEL
ASHRAE Safety Classification	A1	B1	A1	A1	A1	A1	B2
PEL: Permissible Exposure Level; AEL: Acceptable Exposure Limit							

Table 1-3 Chiller Refrigerant Safety Indicators

The above table summarizes the acute and chronic toxicity levels for several common refrigerants. (Note that the information is derived from the MSDSs for the products and other sources believed to be reliable. However, the information is subject to change, must be verified before use, and is not warranted by the sources or by the authors.)

A Note on R-123

An alternative refrigerant used with centrifugal chillers is R-123. This refrigerant is of particular interest regarding toxicity and safety issues.

R-123 is chronically toxic at low levels of exposure: the Acceptable Exposure Limit (AEL) for R-123 is 30 ppm, although some manufacturers recommend a TLV-TWA (Threshold Limit Value-Time-Weighted Average) of no more than 10 ppm.

Although R-123 requires more attention to safety than most refrigerants, it is a safe refrigerant. When the ASHRAE Standard 15-1994 guidelines (discussed later) for handling and storage, and containment and monitoring are practiced carefully, the alternate refrigerants are at least as safe as those they replace. In fact, the EPA reports that with typical new chillers, "the concentration of R-123 can be kept below one ppm" which is well within acceptable levels. In addition, R-123 actually poses less acute risk than R-11, which has been used for decades.

OVERVIEW OF REGULATORY ACTIONS

The Montreal Protocol

Efforts to eliminate Ozone Depleting Chemicals (ODCs) have been going on since the 1970s. Later, international agreements regarding refrigerants were established by the *Montreal Protocol on Substances That Deplete the Ozone Layer*. The Protocol representatives of over 35 nations first met in 1987, coordinated by the United Nations Environment Programme (UNEP); and they established international restrictions on CFC and halon production.

Later meetings (1990 in London, and 1992 in Copenhagen) further accelerated the CFC phaseout schedules and developed a phaseout schedule for HCFCs. Refrigerants that do not contain chlorine (HFCs such as HFC-134a) pose no risk to the ozone layer and their production and use is not scheduled for phaseout. As of spring 1994, the basic Montreal Protocol has been ratified by more than 136 nations. In addition, 88 nations have adopted the amendments agreed to at the London meeting, and 24 nations have adopted the Copenhagen amendments. The Montreal Protocol members also met in October, 1995 in Vienna, Austria.

Montreal Protocol is Accomplishing Its Mission

In a word, the Montreal Protocol is working, and good progress has been made on many fronts since the Montreal Protocol was signed.

- The transition away from using CFCs has had a positive effect on the ozone layer—reports on atmospheric science indicate that the ozone layer will recover faster than previously predicted.

- In the U. S., almost 100% of new car air conditioners are CFC-free, over 90% of foam blowing and cleaning agents used do not contain CFCs, and 20% of the large chillers have been converted to alternative refrigerants or have been replaced with non-CFC equipment.

- For new chillers, the change from CFCs to HCFCs and HFCs is virtually complete. New chillers are primarily using R-123, R-22 and R-134a.[24]

But much remains to be done to complete the job. In the U.S., there are many chillers, automobiles, domestic refrigerators, and commercial and institutional refrigeration and freezing equipment still using CFCs.

Worldwide, a substantial amount of equipment that uses CFCs has yet to be converted to other refrigerants. Of the more that 100,000 chillers worldwide using R-11 and R-12 that are in operation, only about 3,000 have been retrofitted to use R-123 and R-134a.

Future changes to the Montreal Protocol should be approached carefully. We now have time to study the situation and reflect carefully on policy decisions in the future and to make reasoned decisions. All environmental effects, not just ozone depletion, must be carefully considered before we change the protocol. The available technology needs to be evaluated for efficiency and safety as well as its economic impact. The environmental impacts of future actions must be forecast.

Related issues that need attention include[14]:

- The potential for non-compliance to protocol restrictions.

- How developing countries will phase out CFCs.

- Scheduling and curtailing use of methyl bromide.

- Whether the halons currently contained in existing equipment can be withdrawn and destroyed rather than released.

The Clean Air Act

In 1990 in the United States, an amended Clean Air Act was signed into law. These amendments were enacted in accordance with the Montreal Protocol schedules. The EPA, which serves as the "environmental police" in the U.S., is responsible for implementing the Clean Air Act laws.

Title VI of this legislation, entitled *Stratospheric Ozone Protection*, calls for a ban on CFC production. This law states you can continue to use the CFCs you have so long as they are not intentionally released into the environment. But once existing supplies are exhausted, no more will be available.

The 1990 amendment to the United States 1970 Clean Air Act, signed by then President Bush on November 15, 1990 required a production phaseout of ODCs (Ozone Depleting Chemicals) in the U.S. by 2002. In 1992, a Presidential Order accelerated the phaseout of U.S. production to December 1995. The phaseout schedule in the Presidential Order agrees with the phaseout requirements of the Copenhagen agreements.

In addition to listing phaseout requirements, Section 612 of the Clean Air Act Amendment of 1990 requires the EPA to establish a program to identify alternatives to Class I and Class II ODCs and to publish a list of acceptable and unacceptable substitutes. (Class I or II are the specific ozone depleting compounds described in Section 602 of the Act.)[15]

A rule-making effort, which only permits replacement of ODCs with an acceptable substitute, is known as the Significant New Alternatives Policy (SNAP). The EPA published the Advisory Notice of Proposed Rulemaking (ANPRM) for the proposed rule in the Federal Register on January 16, 1992 and published the rule in the Federal Register on June 3, 1993. Under the rule, any substitution of Class I and Class II ODCs must be substances listed as "acceptable" by SNAP. Also, manufacturers, importers, formulators, and processors are required to present information on substitutions through the SNAP ninety-day notification policy.[16]

Federal Excise Taxes

Federal excise taxes on CFCs, based on ODP, will further increase costs.

The Omnibus Reconciliation Acts, passed by Congress in 1989 and 1990, imposed an excise tax on CFCs and other Class I substances. The excise tax was increased in an energy bill signed into law in 1992. This tax on CFCs and other Class I substances is administered by the Internal Revenue Service. (The excise tax does not apply to "feedstock" applications nor to Class I compounds that are recycled.)[17]

These taxes are designed to discourage using ozone-depleting refrigerants by increasing their cost so they are comparable to the prices of the newer, non-ozone-depleting refrigerants. Federal excise taxes on CFCs increased to $4.35 per pound in 1994, to $5.35 in 1995; they will increase to $5.80 per pound in 1996 and will increase an additional $0.45 per pound each year thereafter, to $10.30 in 2006.

It also is possible that additional taxes based on GWP will be levied. This could dramatically increase the costs of R-22 and R-134a which currently have no ODP taxes.

To date, there are no ODP taxes on new refrigerants, but there may soon be a tax based on a refrigerant's TEWI.[18]

Illegal Imports

Several U.S. federal agencies formed a workgroup to investigate claims that CFCs are entering the country illegally.[19]

The workgroup consists of the Environmental Protection Agency's (EPA's) Stratospheric Ozone Protection Division, and Office of Enforcement and Compliance Assurance; the Internal Revenue Service (IRS); and the U.S. Customs Service. The Alliance for Responsible Atmospheric Policy will cooperate in this effort.

To avoid detection, CFC smugglers are shipping in materials falsely labeled as propane or other gases, which places these materials outside the reporting and taxing structures. According to the U. S. Customs, "an importer may obtain a legal permit within EPA's system, import the refrigerant, and ship out the containers filled with sand. He can then report that the refrigerant has been exported, and reuse the paperwork to import more gas."

Table 1-4 below shows how the tax rates for all substances regulated by the Montreal Protocol are increasing. The tax is based on the ODP of the substance[20].

Base Tax, $/ODP pound 1993-1996	
Effective Date	$/ODP pound *
1993	3.35
1994	4.35
1995	5.35
1996	5.80
* The tax is increased $0.45/odp pound in 1996 and beyond.	

Table 1-4   Base Tax, $/ODP pound 1993-1996

The IRS also made clear in early 1995 that all imported refrigerants, including recycled or reclaimed refrigerants, also are subject to the excise tax. Also included are contaminated or "off-specification-listed" ODCs (or "bathtub" blends) created to avoid taxation. This ruling removed the fears of refrigerant manufacturers and wholesalers who were concerned that imports might be purposely contaminated and thereby avoid the tax.

According to Richard A. Kocak, a Treasury Department official, "The person liable for the tax is the person that first sells or uses the ODC after it is entered into the U.S. for consumption, use, or warehousing."

Floor Stocks Tax

One part of the excise tax is called the Floor Stocks Tax. With some exceptions, this tax is imposed on any company that has Class I substances for sale or for use in further manufacturing, or non-refrigeration use, as in the manufacture of insulating foam. Unless the chemical is directly emitted, the tax does not apply to refrigerants used to service an owner's systems, to mixtures (if the mixture contains a non-taxed ingredient that contributes to the product's end use), to feedstock, to recycled Class I compounds, or to final products.

The tax rate is the incremental difference between the tax from the previous year to the current year, and is applied to year-end inventory. In 1994, this tax was applied to:

- 20 pounds or greater of Halons
- 200 pounds or greater of ODCs used in foam insulation
- 400 pounds or greater of all other compounds.[21]

The Climate Change Action Plan

While the majority of current laws and binding regulations focus specifically on ozone-depleting chemicals (CFCs and HCFCs), recent interest has expanded to include other factors.

President Clinton recently issued the Climate Change Action Plan. This sets the goal of reducing U.S. greenhouse gas emissions, which include carbon dioxide, methane, oxides of nitrogen, and HFCs, to their 1990 levels by the year 2000. In this context, refrigerants must be measured in terms of their Total Equivalent Warming Impact (TEWI).[22]

This plan includes a variety of energy efficiency and conservation measures, and targets restrictions on some key chemicals.

The Action Plan directs the EPA and the Department of Energy (DOE) to launch two new coordinated initiatives—Energy Star buildings and Rebuild America—for efficient heating, cooling, and air handling.

The Clinton Administration is adopting a combination of partnership efforts and regulations to minimize the future contribution of HFCs to global warming, without "disrupting the orderly and cost-effective transition away from CFCs."

The EPA will restrict the use and emission of chemicals with high GWP by encouraging cradle-to-grave stewardship for long-lived gases. The EPA will work with manufacturers of long-lived HFCs to get their commitment to not sell chemicals for "emissive use" and to ensure that the gases are captured, recycled or destroyed.

The Climate Change Action Plan is currently a "voluntary" program. If the nation can meet the goals of the program it will remain voluntary. If the goals are not met, mandatory measures will probably be put into effect.

CFC Production Phaseout Schedule

At the 1992 Copenhagen session, the parties to the Montreal Protocol agreed to several amendments that accelerated production phaseout schedules beyond those mandated by the Clean Air Act Amendments. Table 1-5 shows the production levels as called for by the United Nations Environmental Program (UNEP) 1990, the 1990 U.S. Clean Air Act, and the UNEP 1992 amendments.

The amendments call for a total phaseout of CFC production by January 1, 1996 with possible exceptions for essential uses (uses which are necessary for the health or safety of society or are critical for the functioning of society).

The CFCs that will be phased out of production include (but are not limited to) CFC-11, CFC-12, CFC-13, CFC-113, CFC-114, CFC-115, R-500, R-502, and R-503.[23]

Phaseout Schedules

The EPA published a final rule in December 1993 to accelerate the phaseout of ozone depleting substances (ODS) to meet the recent Montreal Protocol changes. For most Class I substances, the new phaseout date is January 1, 1996, with interim cuts.

EPA Schedule for Class I Substances	Allowable Production as % of Baseline* Levels
Year	CFCs
1993	75 %
1994	25 %
1995	(25) % **
1996	0 %

* Baseline year for CFC-22, -12, -113, -114, -115 is 1986
** Attempts were made to go as low as 15% for 1995 and this was the goal, but the percentage for 1995 is not available at this time.

Table 1-5 EPA Schedule for Class I Substances and Allowable Production as % of Baseline Levels

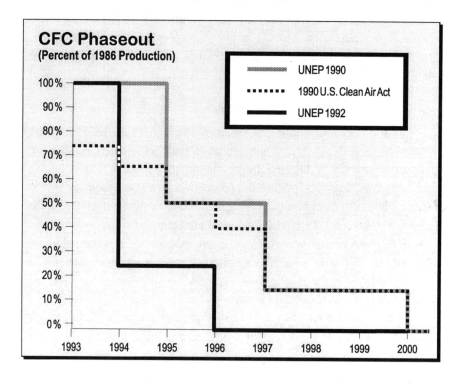

Figure 1-4 CFC Phaseout Schedule

HCFC Production Phaseout Schedule

HCFCs (hydrochlorofluorocarbons) have lower ODP values than CFCs and will be phased out later.

Figure 1-5 below shows the HCFC phaseout schedule recommended by UNEP. The EPA proposes to conform with these recommendations.

Phaseout Schedule for All HCFCs	Allowable Production as % of Baseline Levels
Year	HCFCs
2004	65 %
2010	35 %
2015	10 %
2030	0 %

Table 1-6   Phaseout Schedule for All HCFCs and Allowable Production as % of Baseline Levels

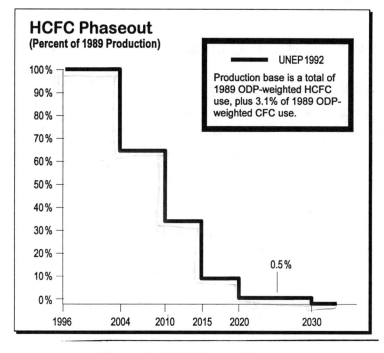

Figure 1-5   HCFC Phaseout Schedule

Proposed U.S. HCFC Phaseout (for air conditioning and refrigeration applications only)		
Year	**Refrigerant**	**Restriction**
2003	HCFC-141b	Production and consumption frozen at baseline levels.
2010	HCFC-22, HCFC-142b	Production and consumption frozen, except for use in equipment manufactured before January 1, 2010.
2015	Other HCFCs	Production and consumption frozen, except for use in equipment manufactured before January 1, 2020.
2020	HCFC-22, HCFC-142b	Production and consumption phased out entirely.
2020	Other HCFCs	Consumption banned for new equipment manufactured on or after January 1, 2020.
2030	Other HCFCs	Production and consumption phased out entirely.

Table 1-7 Proposed U.S. HCFC Phaseout (for air conditioning and refrigeration applications only)

SUMMARY OF KEY REGULATIONS

Overview of Clean Air Act Regulations

The national regulations building owners must comply with are outlined in the Clean Air Act Title VI amendments of 1990. These are rules that have been enacted to bring the U.S. into compliance with the Copenhagen amendments to the Montreal Protocol.

Section 604

Section 604 of the Title VI Clean Air Act of 1990 (effective July 10, 1992) is a final rule that stipulates the phaseout of production and consumption of Class I substances (those with ODPs of 0.2 or greater). This section of the Clean Air Act uses production and consumption allowances to limit the import and production of ozone-depleting substances including CFCs, carbon tetrachloride, halons, and others.

Section 606

Section 606 (effective January 1, 1994) calls for compliance with the accelerated phaseout schedule for Class I and Class II substances.

Section 608

Section 608 of the Clean Air Act amendments of 1990 (enacted July 1, 1992) is the final rule known as the National Recycling and Emission Reduction Program. It mandates compliance with venting bans through requirements to maximize recycling of refrigerants. The intent of this section is to create a supply of refrigerant for use after 1995.

Currently, very little refrigerant is being returned for reclamation, and only a small portion of the certified recovery equipment has been purchased. This leads regulators to suspect that there is significantly more refrigerant leakage occurring than is being reported.

Some fines have been levied, but not nearly as many as could have been. Under Section 608 of the Clean Air Act, "no records" is not an acceptable excuse.

Section 612

Section 612 of the Clean Air Act amendments of 1990 is the Significant New Alternatives Policy (SNAP).

This section of the Act establishes an EPA program to evaluate applications for alternative refrigerants. The EPA's SNAP program provides industry with guidance on acceptable alternatives to ozone-depleting refrigerants. It also lists alternatives it considers "unacceptable" and specifies for which end use the alternatives are banned.

The EPA reviewed and analyzed each applicant. It considers a number of factors including the impact on ozone depletion, global warming, flammability, energy efficiency, and toxicity (worker safety). A very poor score in any one of these categories is likely to mean the refrigerant will be unacceptable, even if it has high scores in all other categories.

Other Sections of the Clean Air Act

Other sections of the Title VI Clean Air Act of 1990 regulate motor vehicle air conditioners, ban the sale or distribution of a number of products that contain CFCs (such as flexible and package foams and aerosol products), mandate labeling of containers, and require that the federal government's procurement regulations conform to the requirements under Title VI.

Regulations Calling for Reduction or Phaseout

The Clean Air Act calls for the reduction or phaseout of various regulated substances that impact refrigeration and air conditioning units. For example:

- All Class I substances are to be phased out by the year 2000—except for methyl chloroform, which is to be phased out by 2002. (Clean Air Act, Section 604)

- CFC production was to be reduced to 25 percent of 1986 levels by January 1, 1994 and to 15 to 25 percent by January 1, 1995. It also is to be reduced to zero by January 1, 1996. (Clean Air Act, Section 606.)

- HCFC-22 is to be phased out for new equipment by January 1, 2010 and is to be completely phased out by 2020. (Clean Air Act, Section 606.)

- HCFC-123 production is to be frozen in 2015, phased out for new equipment in 2020, and completely phased out in 2030. (Clean Air Act, Section 606.)

In addition, the Clean Air Act restricts sales of selected regulated substances. For example, Section 609 places the following restrictions on the sale of CFC-12:

- CFC-12 in containers smaller than 20 pounds can now be sold only to technicians who are certified under the EPA's motor vehicle air conditioning regulations. However, larger containers of CFC-12 may still be sold to people who service appliances other than motor vehicle air conditioners.

- After November 14, 1994 CFC-12 in any size container may be sold only to EPA-certified technicians.

Recovery, Recycling, and Reclamation Regulations

Section 608 of the Clean Air Act governs various aspects of recovery and recycling. For example, it requires:

- Refrigerant recovery and recycling equipment be certified to ensure that emissions are minimized.

- Reclaimed refrigerant to meet ARI Standard 700 for purity. (See "ARI Standards" below for details on these purity standards.)

- Recovery of 80 to 90 percent of the system's refrigerant when working on small appliances such as household refrigerators or freezers. (The specific percentage depends upon the status of the system's compressor.)

In addition, people who service or dispose of air-conditioning and refrigeration equipment must certify to the EPA that they have appropriate recovery or recycling equipment. This certification must be submitted to an EPA regional office.

In the sample certification form on the following page, you will see that establishments with recovery/recycling equipment must indicate the number of vehicles they have at the location. However, more than one vehicle may share the same recovery/recycling equipment.

Also, technicians who work on refrigeration or air-conditioning equipment must use recovery or recycling equipment that has been certified by an EPA-approved equipment testing organization, and they must evacuate the refrigerant to specific vacuum levels. The required vacuum levels depend upon the refrigerant and when the recovery/recycling equipment was manufactured. (Equipment manufactured after November 15, 1993 must be equipped with low-loss fittings and must evacuate refrigerant to somewhat lower vacuum levels than older recycling/recovery equipment.)

THE UNITED STATES ENVIRONMENTAL PROTECTION AGENCY (EPA)
REFRIGERANT RECOVERY OR RECYCLING DEVICE
ACQUISITION CERTIFICATION FORM

EPA regulations require establishments that service or dispose of refrigeration or air conditioning equipment to certify by August 12, 1993 that they have acquired recovery or recycling devices that meet EPA standards for such devices. To certify that you have acquired equipment, please complete this form according to the instructions and mail it to the appropriate EPA Regional Office. **BOTH THE INSTRUCTIONS AND MAILING ADDRESSES CAN BE FOUND ON THE REVERSE SIDE OF THIS FORM.**

PART 1: ESTABLISHMENT INFORMATION

Name of Establishment

Street

(Area Code) Telephone Number

City   State   Zip Code

Number of Service Vehicles Based at Establishment

County

PART 2: REGULATORY CLASSIFICATION

Identify the type of work performed by the establishment. **Check all boxes that apply.**

☐ Type A - Service small appliances
☐ Type B - Service refrigeration or air conditioning equipment other than small appliances
☐ Type C - Dispose of small appliances
☐ Type D - Dispose of refrigeration or air conditioning equipment other than small appliances

PART 3: DEVICE IDENTIFICATION

	Name of Device(s) Manufacturer	Model Number	Year	Serial Number (if any)	Check Box if Self-Contained
1					☐
2					☐
3					☐
4					☐
5					☐
6					☐
7					☐

PART 4: CERTIFICATION SIGNATURE

I certify that the establishment in Part 1 has acquired the refrigerant recovery or recycling device(s) listed in Part 2, that the establishment is complying with Section 608 regulations, and that the information given is true and correct.

Signature of Owner Responsible Officer   Date   Name (Please Print)   Title

-7-

Figure 1-6 Sample EPA Certification Form for Recycling/Reclamation Equipment

The Difference between Recovery, Recycling, and Reclamation

Recovery of refrigerant refers to removing it from an appliance—without necessarily testing or processing it—in order to store, reuse, recycle, reclaim, or transport it.

Recycling and **reclamation** of refrigerant refer to reducing, in some way, the contaminants in used refrigerant. This may include removing moisture, reducing acidity, or removing particulate matter. These procedures may be performed either at the job site or at a service facility. By definition, reclaimed refrigerant meets specific standards for purity (see "ARI Standards" below). Recycled refrigerant does not necessarily need to meet the same rigorous standards that reclaimed refrigerant does. Because of these different levels of required purity:

- Recycled refrigerant can be put back only into the same "system" of machinery belonging to the same owner.

  For example, recycled refrigerant taken from one of several chillers at the owner's location can be moved back and forth among those chillers.

- Reclaimed refrigerant may be sold to a new owner and used in a completely different system of machinery.

The EPA IRG-2 (Industry Recycling Guide), "Handling and Reuse of Refrigerants in the United States," outlines four ways in which recovered refrigerant may be used; specifically, it may be:

- Returned to the original system without recycling it.

- Recycled and reused in the same equipment or other equipment owned by the same person.

- Recycled so that it meets ARI Standard 700 (see below) and used in a different owner's equipment, but the refrigerant stays under the same contractor's custodianship.

- Sent to a certified reclaimer where it will be processed to meet ARI Standard 700 and may be sold to other contractors and used in different owners' equipment.

The EPA had hoped to adopt IRG-2 in early summer of 1995 and establish a final rule governing recycling in the fall of 1995. Note that the reclamation requirements of the section 608 recycling regulations had been scheduled to expire in May of 1995. The requirements have been extended until March 18, 1996, or until the EPA adopts new, more flexible refrigerant purity requirements (based on industry guidelines), whichever comes first. The extension is meant to assure the air-conditioning and refrigeration industry that the EPA is aware of the problems associated with contaminated recycled and reclaimed refrigerants (possible extensive damage to equipment, release of refrigerants, and shortages) and that the EPA never intended for the purity requirements to lapse altogether, but planned to replace them with more flexible requirements.[25]

ARI Standards

Air Conditioning and Refrigeration Institute (ARI) Standard 700-93 defines purity levels for fluorocarbon refrigerants and defines the laboratory analysis methods used to assess the contaminants in the refrigerant. The analysis is performed on a sample from each batch of refrigerant and the contaminant levels can be used to evaluate either new or reclaimed refrigerants. The table below summarizes the general maximum levels permitted for each type of contaminant generally found in refrigerants.

(ARI) Standard 740-93 defines methods for testing and evaluating the performance of refrigerant recovery and recycling equipment. It rates such equipment's performance based on its capacity, speed, purge loss and ability to remove contaminants from refrigerants.

ARI Standard 700-93	
Contaminant	**Maximum Levels**
Acidity	1 ppm by weight
Moisture	10 to 20 ppm by weight *
Boiling Range	Varies with refrigerant
Chloride Ions	Pass defined test
High Boiling Residue	0.01 percent by volume
Particulates/Solids	Pass visual test
Non-condensables	1½ percent by volume
Other Organic Impurities	0.50 percent by weight
	* Varies with refrigerant

Table 1-8 ARI Standard 700-93

Regulations Governing Venting and Leakage

Section 608 of the Clean Air Act regulates the release of specified refrigerants both through venting (the intentional release while servicing, maintenance or disposing of refrigeration or air-conditioning equipment) and through leakage. Specifically, Section 608:

- Bans the venting of CFCs and HCFCs during the service or disposal of refrigeration equipment, except for:
  - Compounds that are not used as refrigerants, such as mixtures of nitrogen and R-22 that are used as holding charges or as leak test gases.
  - Small amounts of refrigerant that are released through the connecting or disconnecting hoses to charge, service, or purge the equipment.
- Requires refrigerant leaks be repaired in equipment with a charge greater than 50 pounds.

The owner must either repair, convert, or replace leaking refrigeration systems or chillers within 30 days if:

- An industrial process or commercial refrigeration system has annual leak rates greater than 35 percent.
- Other chillers (including comfort cooling) have annual leak rates greater than 15 percent.

Fines for non-compliance with the venting ban are up to $25,000 per day of violation. The EPA also will pay a "bounty" of up to $10,000 for information on intentional venting that leads to successful prosecution. These people are serious.

If you install a refrigerant pump-down system, your staff can perform routine maintenance on chillers without using an outside service company and avoid illegal venting. Even though building codes do not require pump-down system use, the EPA does.

Recordkeeping Requirements

Section 608 of the Clean Air Act mandates recordkeeping when CFC and HCFC refrigerants are used. At a minimum, you should have the following information in your files:

- Records of refrigerant purchases.

- A copy of your recovery equipment certification form.
- Records of employee training (re. recycling, handling of refrigerants, etc.).
- Records of recovered refrigerant.
- Facility equipment serviced and the services performed.
- Records of leak detection.
- Records of refrigerant disposition.
- An environmental policy statement.

  This written policy should state that your company and all employees will obey all laws applicable to this industry and that any employee who knowingly vents will be terminated.

Much of the above information can be kept in an ongoing "refrigerant log." A good example of such a log (from Carrier) is found on the next page.

Another recordkeeping requirement is targeted directly at technicians, who must keep proof of their certification at their business office. When they work on equipment that have 50 or more pounds of refrigerant, they must provide owners with invoices that show how much refrigerant they have added to the equipment.

Carrier

Refrigerant Log

Date: _____

Technician: _____

Machine Data

Manufacturer: _____   Refrigerant Type: _____

Model Number: _____   Design Refrigerant Charge: _____

Serial Number: _____

Unit Identification: _____

Location: _____

Leak Identification

Leak Location: _____

Leak Repaired: Yes _____ No _____ Why Not: _____

Leak Detector Used: _____

Status/Comments: _____

Amount of Refrigerant Added: _____

Recovery/Recycle

Equipment Used: _____

Amount Recovered: _____   Recycled: Yes _____ No _____

Re-installed: Yes _____ No _____

Disposition: _____

Status/Comments: _____

Amount of Refrigerant Added: _____

Unintentional Venting

Situation: _____

Approximate Amount Vented: _____

Amount of Refrigerant Added: _____

Summary

Total of refrigerant added as a percent of design refrigerant charge:

General Comments: _____

Figure 1-7   Sample Refrigerant Log from Carrier

Requirements & Guidelines for Technicians

Section 608 of the Clean Air Act requires all HVAC technicians to pass an EPA-approved test given by an EPA-approved certifying organization. As of November 14, 1994 all technicians were to have been certified under the new mandatory program, or they could be "grandfathered"* if they participated in a voluntary program that met most of the EPA's standards.

The deadline for grandfathered technicians to be certified was April 1995. The EPA grandfathered voluntary programs that met *all or most* of the requirements specified for the new mandatory program. The two main HVAC repair programs this affected are those provided by the Air Conditioning Contractors of America-Ferris State University (ACCA-FSU), and the Refrigeration Service Engineers Society (RSES).

Technicians must keep proof of their certification at their place of business. They also must provide their customers with invoices that show how much refrigerant was added to the equipment during the servicing.[26]

Who Must Be Certified

Originally, there was confusion about who was considered a technician and must be certified. Now the EPA has clearly defined a technician as "any person who performs maintenance, service, or repair that could be reasonably expected to release Class I (CFC) or Class II (HCFC) refrigerants from appliances, except for Motor Vehicle ACs, into the atmosphere." This may include installers, contractor employees, in-house service personnel, and owners—anyone who performs the following kinds of activities:

- Connecting hoses and gauges to an appliance to measure pressures.
- Adding refrigerant to an appliance.
- Recovering refrigerant from an appliance.
- Disposing of an appliance (except small appliances) that still contains refrigerant.

---

* *Grandfathered* refers to providing an exemption based on previously existing circumstances. For example, technicians or programs that are grandfathered by the EPA do not need to be re-tested under the new program because of their previous training and test results under a voluntary program.

- Performing any other activity that may "violate the integrity of the refrigerant circuit while there is refrigerant in the appliance."

Persons are not considered to be technicians if their activities are unlikely to interfere with the refrigerant circuit. For example, painting, rewiring external circuits, insulating pipes, tightening nuts and bolts, and other "external" activities would not require a certified technician.

Types of Certification

The EPA defines four types of certification:

- Type I—Servicing small appliances.

- Type II—Servicing or disposing of high-pressure or very high-pressure appliances (except small appliances and motor vehicle air conditioners [MVACs]).

- Type III—Servicing or disposing of low-pressure appliances.

- Type IV—Servicing all types of appliances.

Safety Standards and Regulations

Safety standards and regulations have two main objectives: to protect people and to protect equipment. People need protection from asphyxiation that may be caused by leaks or spills, and from explosions, which may be caused by excessive flammability or pressure. By correctly identifying and labeling refrigerants, especially new compounds, equipment is protected.

Toxicity and Flammability Reference Numbers (ASHRAE 34-1992)

The purpose of Standard 34 (Number Designation and Safety Classification of Refrigerants) is to create an easy, common language for referring to common refrigerants. It includes two major aspects:

- Reference numbers for refrigerants.

- Safety classifications for refrigerants.

The **refrigerant reference numbers** include an optional composition-designating prefix that differentiates between ozone-depleting and chlorine-free compounds (for example, CFC-, HCFC- and HFC-).

The **refrigerant safety classifications** are composed of two characters that indicate the toxicity and flammability for the different classes of compounds.

- Toxicity is indicated by either an A or a B.
  - Class A compounds have a relatively low toxicity rating, with a TLV-TWA (Threshold Limit Value-Time-Weighted Average) of 400 ppm or more.
  - Class B compounds have a higher toxicity rating, with a TLV-TWA less than 400 ppm.
- Flammability is indicated by a 1, 2, or 3.
  - Class 1 compounds are not flammable (no flame propagation).
  - Class 2 compounds have a relatively low flammability rating.
  - Class 3 compounds have a higher flammability rating.

Table 1-9 below summarizes the safety classifications derived from these groups per ASHRAE 34-1992.

ASHRAE Standard 34-1992 Safety Classifications		
	Low Chronic Toxicity (TLV-TWA $\geq$ 400 ppm)	High Chronic Toxicity (TLV-TWA < 400 ppm)
No Flame Propagation	A1	B1
Low Flammability	A2	B2
High Flammability	A3	B3

Table 1-9 ASHRAE Standard 34-1992 Safety Classifications

Safety Code for Mechanical Refrigeration (ASHRAE 15-1994)

All building owners need to be diligent regarding the use and handling of refrigerants. ASHRAE Standard 15-1994 (Safety Code for Mechanical Refrigeration) establishes standards for safety for new refrigerants. The stated purpose of this standard is "... to specify safe design, construction, installation, and operation of refrigerating systems."

Although its specifications are not included in many local codes, in terms of being practical with liability, nothing short of strict compliance with Standard 15 will provide protection. The guidelines set forth in Standard 15-1992 constitute the "industry consensus practice." This doesn't mean they can be ignored. To provide maximum protection against potential lawsuits, strict compliance with Standard 15 is essential.

When ASHRAE 15-1994 Applies

The standards detailed in ASHRAE 15-1994 apply to new systems and when existing systems are modified. For example, they apply when:

- New chillers or heat pumps are installed in a new or existing building.

- New components are added to an existing system.

- Existing parts or components are replaced with ones that provide additional functionality.

- An existing system is modified (converted) to use a refrigerant with a different ASHRAE number designation.

Key Standards from ASHRAE 15-1994

ASHRAE 15-1994 stipulates a wide range of safety measures for equipment and the areas in which it is housed. Some of the major provisions include the following:

- Every equipment room must contain a detector located where refrigerant from a leak would concentrate. When airborne concentrations of refrigerant meet the specified TLV-TWA rating for that refrigerant, the detector should set off an alarm and start mechanical ventilation.

  In some locales, the alarm must be tied to local or central alarm systems. If R-123 is used, some service organizations will not work in a facility until sensors and alarms are installed.

Since the TLV-TWA ratings for compounds are generally low, the detector must be quite sensitive and able to discern relatively small amounts of refrigerant. To avoid "false alarms" due to non-refrigerant fumes (such as from paints or cleaning agents), a refrigerant-specific vapor monitor that is tuned to the specific refrigerant typically is the best choice for this detector.

- Mechanical ventilation is required to prevent build-up of unsafe refrigerant concentrations in some refrigeration rooms.

- The equipment room must have its own ventilation system. The ventilation system serving the occupied space must not be used for this purpose.

- Except for access doors and panels in air ducts and air handler units, there can be no openings that would let escaping refrigerant pass into other parts of the building.

Any access doors and panels in ductwork and air handler units must be gasketed and tight fitting so no refrigerant-contaminated air can flow into the air stream to or from an occupied space.

- If combustion equipment (such as boilers) is used in the machinery room one of two conditions must be met:

  – The equipment room is not the source of the combustion air. (Combustion air must be ducted from elsewhere to ensure that—in case of refrigerant leaks—refrigerant would not be drawn into the combustion chamber.)

  or

  – A refrigerant vapor detector is used to automatically shut down the combustion process if there is a refrigerant leak.

Since shutting down the boilers can be highly inconvenient and costly, it is especially important to avoid "false alarms" due to non-refrigerant fumes. Therefore, a refrigerant-specific vapor monitor that is tuned to the specific refrigerant typically is the best choice for a detector that shuts down the combustion process.

- Up to 330 pounds of refrigerant can be stored in approved containers inside the equipment room. Any additional refrigerant must be stored in vessels that satisfy the "Pressure Vessel Protection" criteria established in Section 9 of the standard.

- Two self-contained breathing apparatus (SCBA) must be placed outside the equipment room—one primary and one back-up unit—in a location where they are readily accessible in case of need.

In addition, this standard includes specifications for the location of refrigeration rooms and other safety guidelines and precautions.

If you want a copy of ANSI/ASHRAE 15-1994, "Safety Code for Mechanical Refrigeration," you can purchase a copy of it ($21 for ASHRAE members, $31 for non-members) from ASHRAE. Call 404-636-8400 for credit card (American Express, MasterCard, Visa or Diners) purchases, or send a check to ASHRAE Publication Sales, 1791 Tullie Circle NE, Atlanta, GA 30329.

Other National Authorities, Standards and Codes

In addition to the Clean Air Act and EPA regulations and the ARI standards discussed above, other professional organizations have established standards, codes, and guidelines that can have an impact on your choice among CFC-reduction alternatives. For example:

- ASME, the American Society of Mechanical Engineers has set forth a refrigeration piping code: B 31.5.

- ANSI-ASTM, the American National Standards Institute, American Society for Testing and Materials, has established material specifications for copper tubing: B-88/B-89.

- Underwriters' Laboratories has established safety standards that have been adopted locally.

- U.L. 2170: *Construction and Operation* and U.L. 2172: *Procedures and Methods:* Underwriters' Laboratories have set safety standards for field conversion or retrofit of products when changing to an alternate refrigerant.

State and Local Regulations

State and local authorities also have enacted regulations covering CFCs. In most cases, even if the local regulation is more strict than the Federal regulations, the local laws apply. Some of the state and local regulations we'll discuss include:

- California Air Quality Management District (AQMD) Rule 1415.

- Local Code Authorities including three major model code bodies.
- Local codes on refrigerant stockpiling.

SCAQMD Rule 1415

California's SCAQMD (South Coast Air Quality Management District) Rule 1415, *Reduction of Refrigerant Emissions from Stationary Refrigeration and Air Conditioning Systems*—is a local rule very much in line with the EPA Clean Air Act. It requires owners to recover, recycle, and reclaim refrigerant.

The purpose of the Rule 1415 is to reduce CFC emissions from stationary refrigeration and air-conditioning systems that contain more than 50 pounds of CFC-based refrigerants. Specifically, it:

- Specifies rules for making, buying, and selling recycled refrigerant.
- Specifies rules regarding recycling and recovery equipment.
- Mandates recycling and recovery of CFCs.
- Bans venting of refrigerants.
- Requires annual inspections for refrigerant leaks.
- Requires recordkeeping and documentation.

In 1995, the SCAQMD enacted new revisions to Rule 1415 that are more stringent and go far beyond the comparable restrictions mandated by the EPA. There are critics of the new revisions who claim they are an unnecessary duplication of EPA's jurisdiction, result in unjustified and excessive costs, and are not reasonable. Some of the key amendments:

- Require additional monthly maintenance and recordkeeping for owners and operators.
- Require repair of *any* refrigerant leaks, regardless of size, within 14 days.
- State that, after July 1, 1995, auditors must be EPA-certified technicians.
- State that anyone buying or selling refrigerants must be an EPA-certified technician.
- Require an annual audit of all CFC systems, including leak testing, summary of leaks, etc.

- Require an annual maintenance program for all HCFC systems, including leak testing and refrigerant use records.

Local Code Authorities

Most local code authorities have not yet changed their specifications to include substitute refrigerants. However, building owners who apply for variances to the existing codes will probably be successful as long as their proposal is consistent with ASHRAE Standard 15-1994 and with the recommendations of model code bodies.

The three major model code bodies: International Conference of Building Officials (ICBO), the Southern Building Code Congress International (SBCCI) and the Building Officials and Code Administrators International (BOCA) all have accepted HCFC-123 and HFC-134a as substitute refrigerants.

In addition, SBCCI and ICBO have adopted ASHRAE Standard 15-1994 on safety practices for all refrigerants.

Many local jurisdictions have established fire and other safety regulations that can be major issues to those who want to stockpile CFC refrigerants for future use.

Managing Your Refrigerant Assets

ASHRAE Guideline 3-1990 provides guidelines for reducing emission of CFC refrigerants in refrigeration and air-conditioning equipment and applications.

This is a manual that gives details on how best to manage your refrigerant assets. It describes the operational parameters that should be monitored at least daily to reduce refrigerant losses. Monitoring of refrigerant systems is made much easier by microprocessor controls.

Recordkeeping: It's up to You...

Many of the rules and regulations the EPA has the authority to enforce require detailed and accurate records. Again we mention the EPA is stepping up their efforts to make sure the regulations are complied with, and facility inspections are part of their function. One way to obey the intent of the law and be prepared for an inspection is to keep good records on *everything* related to your systems that use regulated refrigerants.

The EPA has two types of facility inspections: Level 1 and Level 2.

A *Level 1 inspection* verifies your compliance with the EPA's refrigerant recovery-recycling regulations. An inspector will have a list of items to answer or check off. The inspector has the legal authority to ask you for proof of his requests for information.

If you receive an EPA *Section 114 letter* it usually means that the EPA wants to audit your facility but an inspector is not available, or there was a problem with an earlier inspection, or there was a complaint filed.

This letter contains the same kinds of questions asked in a Level 1 inspection. They usually want similar information, including copies of all venting records. Failure to respond to a Section 114 letter is itself a violation.

A *Level 2 inspection* includes a much longer list of items to answer or check off and involves much closer scrutiny.

If you are given a Level 2 inspection, it might mean that during an earlier Level 1 inspection, you did not have proof that your technician or equipment was certified. A Level 2 inspection also may result from some other situation, possibly a complaint.

To pass a Level 2 inspection, you must have readily available complete, well-organized files for all related information. If your files are in order, it lets the inspector know you are making a good-faith attempt at compliance, and it can save you time and possibly help avoid fines.

An inspector may ask to see the following documentation:

- Equipment invoices and certification forms.
- Invoices, certification papers, and other pertinent data for recovery/ recycling devices.
- Proof that your technicians are certified, which training courses they took and on what dates, certification numbers, etc.

Contractors who maintain, service, or repair air-conditioning and refrigeration equipment must comply carefully with Section 608 of the Clean Air Act (the "venting" rule). To demonstrate that your business is in fact recapturing and recycling refrigerants, you must be able to furnish any information called for in that Section to demonstrate compliance.

Usually the EPA wants to find out:

- The number of jobs the contractor bid in the past three months that involved service, maintenance, repair, or disposal of any appliance or industrial refrigeration unit.

- The amount of refrigerant that was added, recovered, recycled, or reclaimed.

- The kind of recovery and recycling equipment used.

- What was done with refrigerant that was recovered, but wasn't recycled or reclaimed.

- A copy of all invoices for the recovery or recycling equipment used, and documentation on what kinds of refrigerant each piece is designed to be used with.

- A statement with the name, address, and phone number of any off-site reclamation facilities that were used. Also, copies of documents that show a contract with the reclamation facility and how much refrigerant was sent to the facility (including pick-up manifests).

CHAPTER SUMMARY

The use of most common refrigerants is regulated because their release into the atmosphere affects the Earth's environment. Very few will argue that the Earth's ozone layer is not being depleted by CFCs and other substances. Global warming is less understood; scientists will continue to debate how and to what degree it is happening, what causes it, and what the short- and long-term effects are. Be that as it may, the use of any substance that may change the atmosphere and our environment for the worse needs to be examined. What steps to take regarding global warming also are debatable, but we need to take some measures—at least do what we can to minimize refrigerant leaking. With improved refrigerant containment, direct chemical emissions are reduced dramatically and the global warming issues are then addressed by attending to energy-related sources.

The planet we share is getting smaller and is being badly abused. In the following centuries long after we're gone, future generations will be the beneficiaries of what we do or don't do in the near term. Our best hope is to first pass then strictly obey carefully examined regulations.

Whatever the costs and additional effort that result from adhering to regulations, they are well worth it. Not many users of potentially harmful refrigerants are willing to take voluntary responsibility for doing the right thing. If the use of refrigerants is not regulated, many throughout the world will use them carelessly without regard for the consequences. Therefore, all nations must understand and agree that regulations benefit us all.

Environmental protection laws are here to stay, and, as indicated in the graphic below, the number is increasing. As business owners and plant managers, it is your obligation to keep current on the laws. May as well get used to it; new laws generally require more recordkeeping.

An increased awareness about the effects of refrigerants has led to a review of safety standards for both new and old gases. The new standards and codes will help prevent accidents. Understanding and applying these standards, which define an "industry consensus practice," will limit your exposure to liability if someone or something is damaged by an accident arising from a refrigeration system.

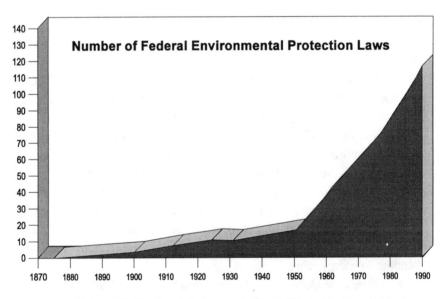

Figure 1-8 Number of Federal Environmental Protection Laws

Chapter 2

Refrigerant Options

Now that CFC production is being phased out, many questions arise about alternative refrigerants. Some of the most common are:

- Are there "drop-in" replacement refrigerants for CFCs?
- Which alternative refrigerant is the best substitute?
- How "permanent" are some of the current replacement refrigerants? What is the future of HFCs and other interim refrigerants considering their impact on global warming?
- Will there be enough supply of CFCs to keep operating the chillers you have?
- How much will refrigerants cost in the future?

This chapter will discuss alternative refrigerant options and try to answer these common questions.

INTRODUCTION

First, let's dispel one common myth about alternative refrigerants. Refrigerant manufacturers have not developed "drop-in" replacements for CFC refrigerants for use in chillers. The fact is, for chiller applications, there are none now. The choices require either modifying existing chillers, which involves changes to heat exchangers, compressors, seals, etc., or purchasing new chillers that use the alternative refrigerants.

Beyond the current group of interim refrigerants, including HFCs, the next generation of new refrigerants will take many years to develop, to test, and to be accepted by the EPA SNAP rule under Section 612 of the Clean Air Act.

However, "near drop-in" HCFC alternative refrigerants do exist for commercial refrigeration applications. Some refrigerant manufacturers have developed interim refrigerant blends that can be used without making any changes to existing machinery (R-408A is one example).

Which alternative refrigerant is best? Of the many alternatives to CFCs identified acceptable by the EPA (from environmental and health perspectives), the best choice will be determined by many factors, including application and efficiency requirements, cost, whether you plan to convert an existing system or install a new one, and more. The refrigerant options are also directly related to decisions regarding machinery options, which will be discussed in the next chapter.

The response by refrigerant manufacturers to the phaseout of CFCs has provided more, not fewer refrigerant choices. The increased selection of refrigerants, along with the many different ways they impact equipment options results in a complex decision confronting many facility operators.

You will probably hear and read conflicting information about which alternative refrigerant is best for a given situation. It's good to remember that equipment manufacturers are naturally biased toward the refrigerant they use in the products they are trying to sell. Consider all of your options, be a knowledgeable consumer, and make an informed decision.

In this chapter we will explore some of the variables that are essential for making this decision, and we will present information on the most common existing alternative refrigerants as well as the new ones being developed. The final choice will be up to you.

Beginning on page 114, there is a series of tables of the most popular refrigerants currently in use (the "Commercial Refrigerants Database"). You may want to refer to these tables as you go through the following pages.

REFRIGERANT VARIABLES

There are many variables to understand and consider when evaluating and selecting a replacement refrigerant:

- The environmental impact including ozone depletion potential (ODP), global warming potential (GWP), and total equivalent warming impact (TEWI) ratings.
- Safety and health issues regarding toxicity and exposure limits, and flammability.
- Chemical behavior.
- Application range (the temperatures at which the refrigerants work best).
- Refrigeration capacity and relative efficiency.
- How long the refrigerant will be available.
- Cost.

Ideally, an alternative refrigerant would have zero or low ODP, low direct GWP, and a level of energy efficiency that either meets or exceeds the refrigerant it is replacing.

When selecting a refrigerant, it is important to consider how well suited the refrigerant is to its application, including operating temperatures, where and how it will be used, and operator expense.

Refrigerant safety is an issue in terms of toxicity and flammability. For any application, new refrigerants should be used safely. Although new refrigerants have been subjected to extensive toxicological testing, any refrigerant can be toxic. Toxic refrigerants may be used as long as they are handled safely. Gasoline is toxic and flammable, but we use it safely in our automobiles; toxic refrigerants are no different.

A BRIEF CHEMISTRY LESSON...

As you consider refrigerant options, understanding some of the basics of refrigerant chemistry may provide some insight into some of the variables we are working with.

In the Periodic Table of Elements, only a small number of elements can be used to synthesize the compounds that can be used as refrigerants. Other compounds are solids, or are toxic, unstable, radioactive, or rare. Some compounds simply will not phase change—that is, they don't react in a way required for a workable refrigerant. (A phase change is defined as a change in state of a substance; for example, a change from a liquid to a gas, or from a gas to a liquid.) This leaves a relatively small number of elements for compounds that can be considered.

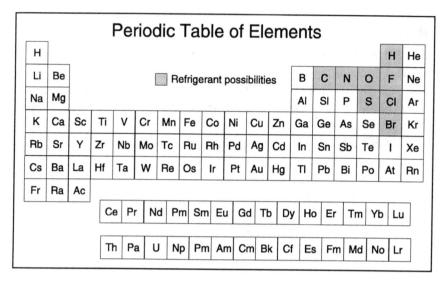

Figure 2-1 Periodic Table of Elements

Refrigerant Chemical Composition

The highlighted area in the table indicates the basic elements we can work with to make refrigerants: chlorine (Cl) and bromine (Br), hydrogen (H), carbon (C), nitrogen (N), oxygen (O), fluorine (F), and sulfur (S). Chlorine and bromine are elements in refrigerants that deplete atmospheric ozone.

Ammonia (NH_3) is made of nitrogen and hydrogen. Sulfur dioxide (SO_2), is made of sulfur and oxygen. These compounds have been used as refrigerants for many years (ammonia is R-717 and sulfur dioxide is R-764), long before CFCs were created.

Both ammonia and sulfur dioxide are considered toxic, and ammonia will burn under certain conditions. But with proper care, ammonia can be used as an industrial refrigerant and recently has been considered for expanded use in commercial applications.

Hydrocarbons, such as methane (CH_4) contain only carbon and hydrogen. The refrigerants of primary interest in this book are the *halocarbon* refrigerants. These are hydrocarbons that have one or more halogens (chlorine, fluorine, or bromine) substituted for hydrogen atoms. (Individual hydrocarbons are also refrigerants. Propane is used widely in industrial refrigeration and has recently become the refrigerant of choice for small air conditioners for some European manufacturers.)

Chlorine derivatives of methane, such as carbon tetrachloride, chloroform, or methylene chloride are used as starting compounds for manufacturing halocarbon refrigerants. The resulting carbon-based (organic) compounds can have many combinations of halogen atoms and hydrogen atoms. To make a refrigerant with two carbon atoms bonded together, you start with a chlorine derivative of ethane (C_2H_6). To make a three-carbon atom refrigerant you start with a chlorine derivative of propane (C_3H_8).

For example, the term CFC (chlorofluorocarbon) applies to a molecule that is fully halogenated and is a chlorinated fluorocarbon, while HCFC (hydrochlorofluorocarbons) applies to molecules containing hydrogen in addition to carbon, chlorine and fluorine. The hydrogen in HCFC tends to destabilize the molecule causing a shorter atmospheric lifetime and less impact on the ozone layer. HFCs have only carbon, fluorine, and hydrogen with no impact on ozone depletion because there is no chlorine. It is the very stable bonds in the CFCs that makes for a long atmospheric life, and that allow time for CFCs to drift to the stratospheric ozone layer where they eventually break apart, releasing the chlorine atom and causing the destruction of ozone.

Most common refrigerants are composed of mainly one or two carbon atoms and many combinations with atoms of chlorine, fluorine, and in some cases, hydrogen. Figure 2-2 below depicts some properties of these elements and compounds made from them.

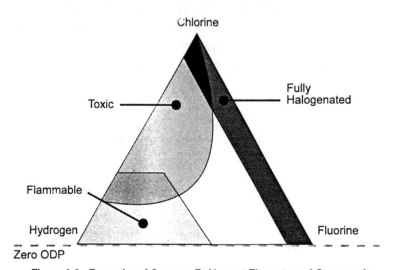

Figure 2-2 Properties of Common Refrigerant Elements and Compounds

Each of the points of the triangle represents a compound consisting only of the element at that point plus carbon. The top of the triangle, for example, represents a compound of chlorine and carbon.

The sides represent combinations of the elements at the adjacent points. The right side of the triangle, for example, represents various combinations of chlorine, fluorine, and carbon. The interior of the triangle represents combinations of all three of the elements plus carbon.

There are many different ways these elements can be combined, and how they are combined will determine whether a given refrigerant is a viable product. As the graphic indicates, refrigerants that contain too much hydrogen are flammable and those that contain too much chlorine tend to be toxic. So that leaves us with the area in the triangle that is not shaded for the best refrigerant choices. Also, refrigerants derived from the compounds along the bottom of the triangle do not contain chlorine and will have zero ODP.

Figure 2-3 shows the CFC derivatives of methane—these compounds have only one carbon atom. The CFCs that are being phased out (R-11 and R-12) are along the side of the triangle that have the characteristic of being fully halogenated; that is, they have a long atmospheric life.

Figure 2-4 shows the CFC derivatives of ethane—these compounds have two carbon atoms. The refrigerants R-123 and R-134a are in this family of compounds. Both are in the area of the triangle where the best alternatives are found.

In the bottom row of the graphic are compounds that do not contain chlorine, and have zero ODP. The common compounds used in refrigerants "blends" are indicated in the graphic. (Blends are refrigerants composed of more than one component and are discussed in detail later in this chapter.)

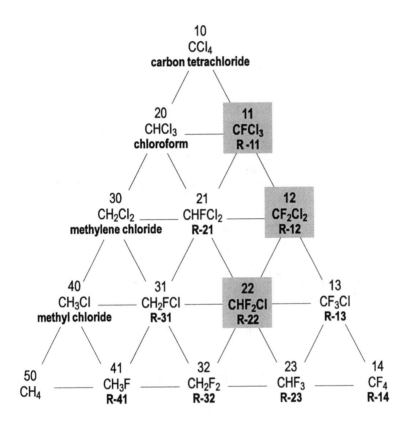

Figure 2-3 CFC Derivatives of Methane

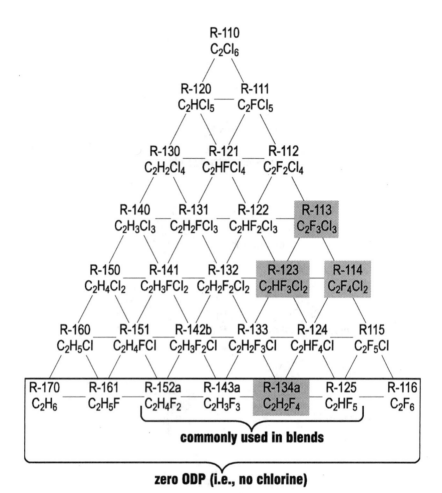

Figure 2-4 CFC Derivatives of Ethane

The next graphic in our brief chemistry lesson shows the relative positions of the most common refrigerants and their relative levels of toxicity, flammability, and ODP.

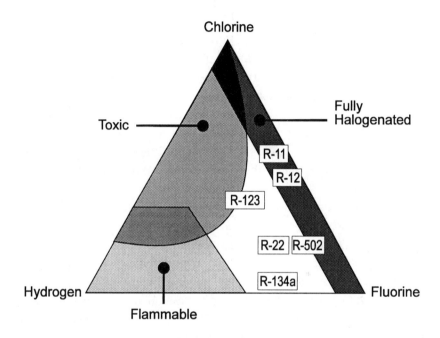

Figure 2-5 Relative Levels of Toxicity, Flammability, and ODP
for Common Refrigerants

Another important aspect of the chemistry of refrigerants is the relative size of the molecules. R-22 for example, has a smaller molecular weight than R-11 or R-123. The molecular weight of the refrigerant has a direct bearing on the size of the equipment in which the refrigerant is used.

Different combinations of elements affect both thermodynamic and thermophysical properties, which in turn determine the capacity, efficiency, and application temperature range of a particular refrigerant.

Naming the Refrigerants

You may wonder how the names and numbers of refrigerants are determined, and what the numbers and the letters in the refrigerant names mean. Both the letters and numbers indicate the compounds that make up the refrigerant.

Refrigerant Numbering

The chemical composition of a refrigerant determines the refrigerant number. As refrigerants become viable commercial products, the American Society of Heating, Refrigeration, and Air-Conditioning Engineers (ASHRAE) assigns the numbers used to identify refrigerants. These numbers provide a shorthand method for determining the chemical composition of refrigerant compounds.

ASHRAE Standard 34-1992 describes (in about a dozen pages) the details of the numbering system, which, with some knowledge of chemistry, can be understood in its entirety. The following few paragraphs review the basics of Standard 34, and explain some of the more common notations.

For pure refrigerants related to methane (only one carbon atom), to ethane (two carbon atoms), and, for most cases those related to propane (three carbon atoms), the molecular structure can be determined by the refrigerant number. The table on the next page indicates to which numbering series each of these belong.

The zeotropic and azeotropic refrigerant blends (which are discussed in detail in a later section in this chapter) are designated by a 400 or 500 number, respectively. These are assigned successive numbers as the refrigerants become commercially available. Inorganic compounds (those without carbon) such as ammonia, have a 700 number that is determined by adding the molecular weight of the compound to 700. (For example, ammonia, or R-717 is an inorganic compound of the 700 series. The "17" is derived from NH_3 where the atomic weight of Nitrogen is 14 and Hydrogen is 1. This gives us $14 + (1 \times 3) = 17$.) The sum of the atomic weights gives the molecular weight.

For hydrocarbons and halocarbons, the right-hand-most number is the total of fluorine (F) atoms in the compound. The second number from the right is one more than the number of hydrogen (H) atoms. And the third number from the right is one less than the number of carbon (C) atoms, which is omitted if it equals zero.

Using these rules for R-22, we can determine the composition of this refrigerant's molecules: the right-hand-most number equals the number

ASHRAE Standard 34-1992 Refrigerant Numbers	
000 series	Methane-based compounds
100 series	Ethane-based compounds
200 series	Propane-based compounds
300 series	Cyclic organic compounds
400 series	Zeotropes
500 series	Azeotropes
600 series	Organic compounds
700 series	Inorganic compounds
1000 series	Unsaturated Organic compounds

Table 2-1   ASHRAE Standard 34-1992 Refrigerant Numbers

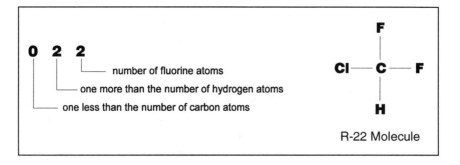

Figure 2-6   R-22 Molecule

of fluorine atoms, which is 2; the second number from the right minus one equals the number of hydrogen atoms (2-1=1); and the first number is zero because it is one less than the number of carbon atoms (1-1=0).

The number of chlorine (C_l) atoms in the compound can be determined by adding all the atoms that can be connected to the carbon atoms and subtracting the number of fluorine and hydrogen atoms. With R-22, there are four atoms connected to one bonded carbon atom. Since there are two fluorine atoms and one hydrogen atom, this would leave room for one chlorine atom.

Now, with R-123, we can again determine the composition of this refrigerant's molecules: the right-hand-most number equals the number of fluorine atoms, which is 3; the second number from the right minus one equals the number of hydrogen atoms (2-1=1); and the first number plus one equals the number of carbon atoms (1+1=2).

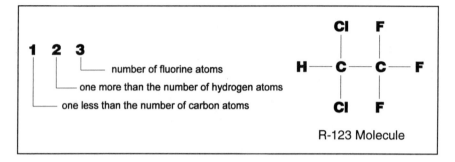

Figure 2-7 R-123 Molecule

Once again, the number of chlorine (Cl) atoms in the compound can be determined by adding all the atoms that can be connected to the carbon atoms and subtracting the number of fluorine and hydrogen atoms. With R-123, there are a total of six atoms connected to the two bonded carbon atoms. Since there are three fluorine atoms and one hydrogen atom, this would leave room for two chlorine atoms.

Now 134a has a lower case "a" after the number and 245ca has two lower case letters, "ca." These lower case letters indicate the arrangement of the atoms attached to the carbon atoms. The term *isomer* is used to indicate that there are various ways to arrange the atoms (isomers have the same number and kinds of atoms, but in different arrangements).

For example, 134:

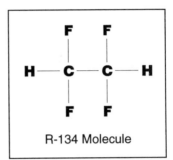

Figure 2-8 R-134 Molecule

and 134a:

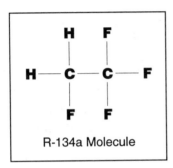

Figure 2-9 R-134a Molecule

R-134 and R-134a both have the same number of fluorine and hydrogen atoms, but in different arrangements. The more symmetrical molecule is indicated by the number only.

The less symmetrical molecule is indicated by adding a lower case "a." The symmetry is determined by subtracting the sum of the atomic weights of the atoms attached to one carbon atom from the sum of the atomic weights of the atoms attached to the other carbon atom. The smaller absolute value of the difference indicates the more symmetrical isomer.

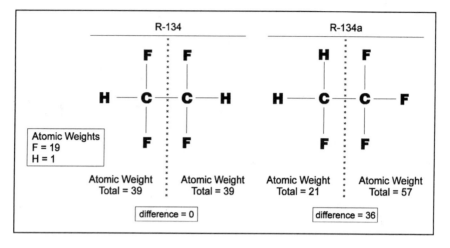

Figure 2-10 Molecule Symmetry

For the propane series, since there are three carbon atoms, there are even more possible combinations. In this case, the first lower case letter indicates the arrangement on the center carbon atom and the second letter indicates the symmetry of the atoms around the first and third carbon atoms. An example is 245ca.

Composition-Designating Prefixes

The letters used to designate refrigerant composition indicate whether chlorine, hydrogen, or fluorine are present. The capital letter "C", for carbon, is preceded by the letters C for chlorine, F for fluorine, or H for hydrogen, as in CFC, chloro-fluoro-carbon or HFC, hydro-fluoro-carbon. These composition-designating prefixes are used primarily in non-technical publications where ozone depletion is of interest.

Otherwise, the more technically correct way to designate refrigerants is to put the upper case letter "R" in front of the refrigerant number to designate that the compound is a refrigerant, such as R-134a. Blends such as R-507, also can be written as R-125/143a (45/55) which gives the components of the blend (R-125 and R-143a) and their percentages (45% and 55%).

Safety Classifications

Although not indicated in their names, refrigerants also are classified into safety groups by using a combination of the letters A and B (for toxicity) and the numbers 1, 2 and 3 (for flammability).

Class A signifies refrigerants which are not toxic at concentrations less than or equal to 400 ppm based on a time-weighted average.

Class B refrigerants are toxic at levels below 400 ppm.

Class 1 indicates no flame propagation.

Class 2 has a low flammability limit.

Class 3 are highly flammable.

In the case of blends, there are two safety classifications — first, for the blend as it is formulated and second, for the worst-case "fractionation" (fractionation, simply put, is a change in composition primarily due to refrigerant leaking).

ASHRAE Standard 34-1992 Safety Classifications		
	Low Chronic Toxicity (TLV-TWA ≥ 400 ppm)	**High Chronic Toxicity (TLV-TWA < 400 ppm)**
No Flame Propagation	A1	B1
Low Flammability	A2	B2
High Flammability	A3	B3

Table 2-2 ASHRAE Standard 34-1992 Safety Classifications

Examples of Each Safety Classification	
A1	Nearly all commercial refrigerants incl. R-12, R-502, R-22, R-134a, etc.
A2	R-152a, R-142b
A3	R-290 (propane)
B1	R-123
B2	R-717 (ammonia)
B3	Vinyl chloride

Table 2-3 Examples of Each Safety Classification

PURE REFRIGERANTS AND BLENDS

Refrigerants can be classified as pure fluids or as mixtures or blends.

A Pure Fluid

A pure fluid refrigerant is chemically made of one component of a single kind of molecule. At a given pressure, the temperature of a pure refrigerant does not change when it boils or condenses in the refrigeration cycle. Some examples of pure fluids are R-12, R-22, and R-134a.

A Mixture or Blend

The terms *mixture* and *blend* are interchangeable and both describe refrigerants composed of more than one component or kind of molecule.

The components used to make refrigerant blends are selected specifically to create a final product with specific characteristics. These characteristics, such as capacity, efficiency, discharge temperature, vapor pressure, etc., will vary depending on the percentages of the components that make up the blend.

Zeotropes and *azeotropes* are two types of blends. Some use a third classification; the Near-Azeotropic Refrigerant Mixture, or NARM. But all NARM refrigerants are more precisely defined as zeotropes, because they must be handled in the same way as zeotropes. It looks as if the NARM category is not going to gain wide acceptance.

As we discussed earlier, one way to distinguish zeotropes from azeotropes is by the ASHRAE number series given to these blends. Zeotropes are the 400 series and azeotropes are the 500 series.

Azeotropes

An azeotrope is a mixture of two or more liquids that boils at a constant temperature.

An azeotrope behaves basically like a single fluid. At a given pressure, the temperature remains constant as the refrigerant boils or condenses. This temperature can be either higher or lower than the boiling temperature of any one of the individual component liquids.

An azeotrope, at a specific pressure, does not change composition—that is, the components remain blended and don't separate when the blend evaporates or condenses. This is because at a certain pressure the combination of components have one boiling temperature—its "azeotropic point." (It may not behave this way at other pressures, but the differences are very slight.)

The most common azeotrope in wide use is R-502, a blend of R-22 (48.8%) and R-115 (51.2%) [typically written R-22/115 (49/51)]. For all practical purposes, R-502 performs like a single fluid even though the R-22 and R-115 are not chemically joined, only mixed together.

Zeotropes

A zeotrope mixture is a blend with two or more components that, at a constant pressure, exhibits a distinct and substantial shift in temperature during condensing or boiling. The components in a zeotrope mixture do not have a constant boiling point (and therefore, there is no constant temperature of all of the components taken as a whole).

This temperature change during a constant pressure phase change is called *glide* (see below), and varies with the components used in the blend. The amount of temperature glide (or boiling range) for a particular zeotrope is a measure of its deviation from being an azeotrope.

Some of the more common zeotropes that exhibit high glide include:

- R-401A, a popular interim R-12 replacement which has an 11° F glide (boiling range).

- R-407C, an R-22 replacement refrigerant, which has from 8 to 12° F glide.

Experience with zeotropes as replacements for R-12, R-22, and R-502 is growing, and all major refrigerant manufacturers now produce refrigerants with glide. Competent technicians find they are easy to use and require handling and servicing techniques similar to those commonly used with traditional refrigerants.

Glide

Glide is a temperature change during a constant pressure phase change. (Again, a phase change is defined as a change in state of a substance; for example, a change from a liquid to a gas, or from a gas to a liquid.) Glide is a function of refrigerant blends; the boiling temperature of "pure fluid" refrigerants does not change, and therefore this type does not have glide. The amount of glide (or the boiling temperature range) varies with the components used in a blend.

Refrigerant glide can be explained in terms of the nature or properties of the refrigerant (which is temperature glide, or glide due to a phase composition shift), and also in terms of glide caused by a drop in pressure in a heat exchanger and other "system" influences (system performance glide).

"Temperature glide" is caused by the effects of one refrigerant component evaporating or condensing before another component at a different temperature and pressure. This is glide due to a composition shift during a phase change (refer to the graphic on page 80).

In an operating refrigeration system, the boiling range characteristic of a refrigerant blend causes different temperatures between the inlet and outlet locations of a direct-expansion evaporator or condenser. These temperature differences in the heat exchangers of the system is temperature glide. (The glide value for a particular refrigerant may be expressed as a range because it can be measured in different ways and based on several variables. It also may vary depending on operating temperatures. Also, in the evaporator, temperature glide will be smaller than the boiling range because the refrigerant will enter the evaporator partially vaporized.)

Temperature glide is primarily observed as a characteristic of zeotropic blends. However, temperature glide also does occur with near-azeotropic and azeotropic blends, but to a lesser extent than with zeotropes. Since azeotropes do not have temperature glide at only one pressure/temperature relationship, they are considered low-glide blends at other conditions.

"System performance glide" is a change in temperatures caused by either the effects of a drop in pressure in a heat exchanger (a pressure differential caused by friction), or the effects of solubility of oil. All systems use special oils for lubrication, mostly in the compressor. Some of the oil travels with the refrigerant through the system and the presence of the oil can cause a very slight glide effect.

For some refrigerants, the "system performance" effects may be greater than the effects of "temperature glide." So to determine the total glide for an actual system, both aspects must be considered.

As an example, a refrigerant may experience only 0.1° F glide as a function of the properties of the refrigerant, but the effects of a pressure drop and the presence of oil could be as much as 2° F glide. This means the system affects total glide more than the refrigerant itself does.

Other examples are provided in the table below.

R-22 and R-407 Glide			
	System Glide*	Property Glide **	Total Glide
R-22 (theoretical)	0° F	0° F	0° F
R-22 (in actuality)	2° F	0° F	2° F
R-407C (in actuality)	2° F	10° F	12° F
* Glide due to pressure differential ** Glide due to refrigerant properties			

Table 2-4   R-22 and R-407 Glide

The nature of refrigerants with glide can provide opportunities for improving system energy efficiency. This can be done by changing the design of cooling coils to match the change in temperature of the refrigerant inside a heat exchanger to the change in temperature of the external fluid (the chilled water or air). The design of cooling coils is currently the subject of research, especially in the area of counterflow heat exchangers. (See "Evaporator Flow Design" in Chapter 3.)

An Example of Glide Using R-407C

Let's look at temperature glide using the zeotrope R-407C as an example. R-407C is a blend of R-32/125/134a (23/25/52) and has a 12° F glide or boiling range.

Each of the components of a refrigerant blend has a different relative *volatility,* which is to say they boil and evaporate (change states) at different temperatures. As the liquid exits the expansion valve and begins to vaporize (see Figure 2-11) the vapor in the evaporator consists of a portion of all three blend components. The vapor is, however, richer in the most-volatile component (R-32 in this example) because it is boiled off at a greater rate than the other components with higher boiling points.

This means that the composition of the remaining liquid shifts as the liquid boils and turns to vapor. And as the composition shifts, the "bubble point" of the remaining liquid rises. (More on the subject of bubble point follows this example.) What remains is a liquid richer in the less-volatile components (R-125, and R-134a in this example). The composition of the liquid continues to shift as the liquid turns to vapor, until the liquid reaches its maximum boiling point.

The diagram in Figure 2-11 demonstrates that as the boiling point rises, the composition and temperature of the refrigerant gas changes. This temperature glide is caused by the changing proportions of the individual components that are vapor and liquid.

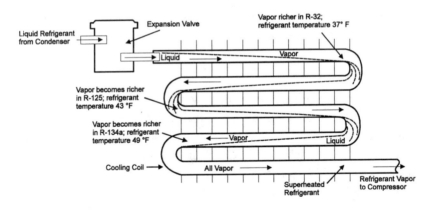

Figure 2-11   Example of Temperature Glide

Eventually the last drop of liquid boils and the rest of the refrigerant is vaporized and the refrigerant continues to the compressor and on through the rest of the refrigeration cycle. (In some applications, as in a grocery store refrigeration system, additional heat might be picked up after the refrigerant has fully evaporated in the long refrigerant lines between the evaporator and compressor. This produces superheated refrigerant.)

Temperature Glide in Terms of Bubble Point and Dew Point

Another way to look at temperature glide is in terms of a refrigerant's bubble point and dew point.

Any given refrigerant blend has a boiling point for a given pressure at which vapor starts to form when heat is added. This is the "bubble point" for that composition and pressure.

The bubble point usually is found (but not always) between the boiling points of the individual pure components of the blend.

As noted earlier, R-407C is blend of 23% R-32, 25% R-125, and 52% R-134a. R-32 has an atmospheric boiling point of -61° F, R-125 a boiling point of -55.4° F, and R-134a -15° F; but the bubble point of the blend as a whole is -48° F. (These temperatures are atmospheric boiling points. The actual refrigerant boiling temperatures in a heat exchanger depends on the pressure it is under.)

Once the liquid mixture reaches the bubble point, it starts to boil as heat is added, and vapor forms.

The composition of the remaining liquid changes during the boiling process. As the composition shifts, the "bubble point" of the remaining liquid rises.

As the last drop of liquid boils (vaporizes) inside the cooling coil (the evaporator), the evaporation temperature has risen to a temperature called the "dew point."

The "boiling range" is the range of temperatures between the bubble point and the dew point and is equivalent to the temperature glide.

Here's an example that shows how, with a zeotropic refrigerant blend, the composition of the refrigerant and the refrigerant temperature both change as the refrigerant progresses through the evaporator and the condenser.

Using a Pressure-Enthalpy diagram (Figure 2-12), we see that the evaporator inlet (D) and outlet (A) temperatures are different, and the condenser inlet (B) and outlet (C) temperatures are different. At all other points in the system the fluid behaves normally. (Enthalpy [or internal energy] measured on the x-axis in the graphic below is the energy content measured in British Thermal Units (BTUs) per pound of refrigerant.)

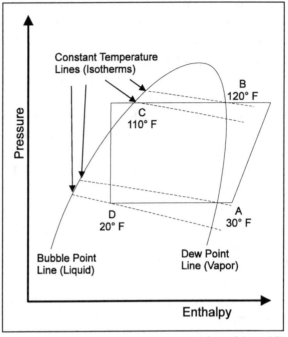

Adapted from ICI

Figure 2-12   Temperature Glide in Terms of Bubble Point and Dew Point

In Figure 2-12, which shows typical system performance with a zeotropic refrigerant blend, the flow of the refrigerant through the entire refrigeration cycle is as follows:

A to B: through the compressor.

B to C: through the condenser.

C to D: through the metering device (expansion valve).

D to A: through the cooling coil (evaporator).

Fractionation

A leak can occur in any air-conditioning or refrigeration system. Refrigerant vapor can leak out, and air can leak in. Of particular concern regarding refrigerant blends is the potential for a change in composition of the mixture when leaks occur. This is called *fractionation*.

As we have seen, a characteristic of a refrigerant blend is that the components that it is made of vaporize and condense at different temperatures and the parts of the refrigerant that are liquid and vapor experience a composition shift. As a result, at certain times the vapor consists of a greater percentage of one component than it does of the others.

So if a leak happens in any part of the system where both liquid and vapor refrigerant are present, as in the evaporator or condenser, only the components that are vapor will leak out of the system. The remaining refrigerant liquid or vapor then becomes richer in one component than in others.

This change in the percentages of the blend compounds remaining in the system is called fractionation. Since refrigerant blends are specifically composed of certain mixtures of compounds, this change can reduce system capacity and efficiency.

The greatest concern for fractionation is the case where the remaining mixture becomes flammable, especially in systems not normally designed to handle flammable refrigerants. (Leak detection equipment is used in situations where flammability caused by fractionation is an issue.)

Note that most leaks do not occur in an area of the system where phase changes happen—where there is both liquid and vapor. They are more likely to occur in a "gas" suction line or a "liquid" line. In these cases, the refrigerant does not experience fractionation because any refrigerant that leaks is of the same composition of the original refrigerant charge.

After fractionation occurs, the chemical composition of the refrigerant must be analyzed using charge/recharge tests. The composition typically cannot be measured on-site. A sample of the refrigerant needs to be sent to a lab for testing to determine the composition. However, laboratory testing has shown that the effects of leaking or fractionation of most blends are only minimal for most applications, even with several repeated leak and recharge cycles. In cases where the proportions are significantly altered and the effects are not minimal, the entire refrigerant charge must be replaced.

System Charging and Fractionation

Another possible way fractionation can occur with systems that use a zeotrope refrigerant is by refilling the system (charging) with vapor instead of with liquid refrigerant. Since the refrigerant can change phase from liquid to vapor inside the refrigerant container cylinder, one component again will vaporize before the others and the system will be charged with more of that component. So to make sure the composition stays balanced, all zeotrope refrigerants must be charged with liquid refrigerant from the cylinder. ("Pure" refrigerants or azeotropes do not have this restriction; systems that use these can be charged with either vapor or liquid.)

Most charging cylinders that contain blended refrigerants have dip tubes that allow technicians to charge a system with liquid with the charging cylinder in an upright position.

If it is necessary to charge refrigerant into the low-pressure side of a system (before the compressor), do not vapor-charge from the cylinder. Instead, vaporize the liquid refrigerant coming from the cylinder using an expansion device or vaporizer. This will prevent damage to the compressor caused by "slugging" due to the presence of liquid. Refrigerant can be admitted to the high-pressure side of the system (after the compressor) as either a liquid or gas.

Refrigerant manufacturer AlliedSignal has said that a system charged with R-404A that has experienced fractionation can be serviced with the blend R-507. R-404A has R-125 as one component and also 4% R-134a. The R-125 component will leak out first because it has the lowest boiling point.

The blend R-507 has R-125 as one component but does not have any R-134a. Servicing R-404A with R-507 will reduce the proportion of the R-134a in the mixture but will replace the lost R-125 and return the refrigerant in the system more closely to the composition of the original charge.

In spite of this observation, mixing blends in this way is experimental and you should always seek expert advice when servicing your chiller and refrigeration equipment.

Influence of Leak/Recharge on Performance

When fractionation occurs, the composition of the remaining refrigerant has changed and the system needs to be recharged. Technicians may not be able to simply "top-off" a refrigerant charge as is the current practice with "pure" refrigerants. If a small amount of fractionation has occurred, (a lab test sample can indicate how much fractionation has occurred) topping off the system may be acceptable. Otherwise all refrigerant must be evacuated from the system and the system must be recharged.

If there is an R-407C leak from an operating unit in an area where both liquid and vapor are present at the same time, as in the heat exchangers or after an expansion device, both vapor and liquid will leak from the unit. The composition of the refrigerant left in the system will remain essentially unchanged from the original composition. After completely recharging the system with R-407C, the performance of the unit will be the same as its original performance with R-407C.

However, if the unit is not operating and a vapor leak causes fractionation, the refrigerant remaining in the unit will not be of the original composition and performance will be affected. The influence of fractionation on the performance of R-407C is summarized in the table below.

The table assumes that the system leaks 50% of the charge out and is then refilled to capacity (topped-off) with the original blend. The question is, What impact does repeated refilling have? The results show after repeated refilling there is some lost capacity and the refrigerant cannot remove heat as efficiently.

Theoretical Unit Performance After 50 Percent Weight Vapor Leaks and Recharges of R-407C				
Recharge #	Rel. COP*	Rel. Cap.**	Compressor Discharge	
			Temperature	Pressure
0	100%	100%	178°F (81.1°C)	261 psig (19.03 bar, 1903 kPa)
1	101%	95%	177°F (80.6°C)	246 psig (18.00 bar, 1800 kPa)
2	101%	93%	177°F (80.6°C)	239 psig (17.51 bar, 1751 kPa)
3	101%	92%	177°F (80.6°C)	236 psig (17.31 bar, 1731 kPa)
4	101%	91%	177°F (80.6°C)	235 psig (17.24 bar, 1724 kPa)
5	101%	91%	177°F (80.6°C)	235 psig (17.24 bar, 1724 kPa)

* Coefficient of Performance (measure of energy efficiency) relative to the Coefficient of Performance of the original charge of R-407C.

** Refrigerant cooling capacity relative to the capacity of the original charge of R-407C.

DuPont

Table 2-5   Theoretical Unit Performance After 50 Percent Weight Vapor Leaks and Recharges of R-407C

EVALUATING ALTERNATIVE REFRIGERANTS

There is a lot of refrigerant testing going on as the search continues for the best alternatives for the many different kinds of applications.

The most common CFCs that are being phased out are R-11, R-12, and R-502. At this time, the most prevalent alternatives for these refrigerants are R-123, R-22, and R-134a, but other alternatives are being evaluated. R-22, a possible replacement for both R-12 and R-502, will also be phased out eventually and a long list of replacement refrigerants for R-22 are being tested and evaluated.

"Interim" refrigerants and blends are mostly those in the HCFC family. These contain chlorine and will eventually be phased out, but can be manufactured and used until the year 2030 (tables of phaseout dates are provided in Chapter 1). "Long-term" solutions are mostly those refrigerants (including blends) that do not contain any chlorine, such as the HFCs.

Basically, there are two kinds of refrigerant testing: theoretical testing and testing in actual applications. As might be expected, the way refrigerants behave and perform in theory or through simulations can be very different from the performance of refrigerants tested in actual applications and systems.

Typically, theoretical testing and comparisons are made using simulation computer programs that are given data from compressor and heat transfer tests. These simulations assume zero superheat. (Superheat is heat added after the refrigerant has fully evaporated. For example, the refrigerant vapor carried in the long lines between the evaporator and the compressor in grocery stores tend to pick up additional heat.)

In addition, the theoretical comparisons do not consider that compressor pumping efficiency can vary between refrigerants, and for refrigerants with glide, the assumptions made in the simulations about condenser and evaporator performance can have a large impact on overall performance results. For example, in the simulations, no heat exchanger surfaces are actually used, and performance based on counterflow design in the evaporator or condenser is not considered.

To completely and accurately assess the performance of a given refrigerant requires testing in typical existing designs or in new, fully optimized system configurations, depending on the intended use of the refrigerant. This testing in actual applications is time consuming and expensive.

To narrow down the field of potential substitutes, relative performance comparisons have been made using compressor calorimeter tests, and computer modeling. Instead of actual systems, so-called "soft-optimized" systems are being used for testing until actual systems are built.

All of the candidate refrigerants being tested have issues that must be resolved before they can be used successfully. Two major issues are: the design changes necessary to compensate for the variation in capacity and pressure of different refrigerants; and the design changes necessary to accommodate the potential for composition shifting or glide.

As stated earlier, the results of a theoretical comparison can be quite different from the results produced in an actual system. A theoretical comparison does not consider the following actual application system behavior or attributes:

- Suction line superheat can be quite high in a commercial system. For example, a low temperature supermarket refrigeration system can have 50 degrees of non-productive suction.

- Discharge and suction line pressure drop.

- Differences in liquid and vapor specific heat (the amount of energy required to raise the temperature of one pound of a refrigerant one degree Fahrenheit, usually expressed as Btu/lb. ° F) which leads to differences in the performance of heat exchangers and subcooling systems (subcooling refers to the additional "cooling" of liquid by outside air or an external refrigeration system). For example, liquid subcooling with R-22 can add 20% additional capacity compared to 40% additional capacity with R-404A.

In Chapter 4, the test results from a lab using actual systems designed to accommodate and adjust for the new refrigerants are presented as well as the results from other tests.

EPA-Accepted Refrigerant Substitutions (SNAP)

To some extent, evaluations for alternatives have been completed.

Part of the Clean Air Act required that the EPA establish a program to identify alternatives to refrigerants that contain ozone-depleting compounds, and to publish a list of acceptable and unacceptable substitutes.

The EPA has responded by designating a list of approved end uses for refrigerants that are being phased out in Section 612 of the Clean

Air Act amendments of 1990, the Significant New Alternatives Policy (SNAP). Under this rule, any substitution of Class I and Class II ODCs must be substances listed as "acceptable" by SNAP.

The refrigerants listed below are only three of many that the EPA has designated approved end uses for under SNAP. From time to time there are additional new refrigerant candidates under consideration.

And note that this is not a comprehensive portrayal of all options available and it only represents the EPA's conclusions to date. Although the designation as "acceptable end uses" by the EPA for these refrigerants is based on strict adherence to the recommendations of ASHRAE Standards 15 and 34, the recommendations are not entirely accurate or completely up-to-date. For example, the EPA says you are allowed to use R-22 in R-12 machinery. In reality, this direct replacement requires changes to the system compressor and piping and is usually not possible or practical.

The EPA's SNAP list only begins to scratch the surface regarding alternative refrigerants. Other refrigerants that are not included in the list are currently being applied, and some examples are discussed later in this chapter. Please don't base any business decisions on what's contained in the list, since there are many different issues to consider.

Note also that although the EPA lists acceptable end uses for R-22, it does not list substitutes for this refrigerant because it is being phased out after year 2000 (the current EPA schedule calls for a phaseout of R-22 for new equipment applications by the year 2010). At this time, according to the EPA, any substitute is acceptable for R-22. (Specific alternatives for R-22 are discussed in a later section in this chapter.)

According to the EPA[27]:

- HCFC-123 is acceptable as a substitute for:
    - CFC-11 in centrifugal chillers, both in new equipment and in retrofits.
    - CFC-12 and CFC-500 in new centrifugal chillers.
- HCFC-22 is acceptable for use in *new* equipment as a substitute for:
    - CFC-11 in centrifugal chillers.
    - CFC-12 in centrifugal chillers, reciprocating chillers, cold storage warehouses, residential dehumidifiers, residential freezers, commercial ice machines, industrial process refrigeration equip-

ment, refrigerated transport equipment, retail food systems, vending machines, and water coolers.

- CFC-500 in centrifugal chillers, dehumidifiers and refrigerated transport systems.

- CFC-502 in cold storage warehouses, residential freezers, commercial ice machines, industrial process refrigeration systems, refrigerated transport systems, and retail food systems.

• HCFC-22 is acceptable for use in *existing* equipment, or retrofits, as an alternative for:

- CFC-12 in cold storage warehouses, industrial process refrigeration equipment, retail food systems, and vending machines.

- CFC-502 in cold storage warehouses, industrial process refrigeration equipment, retail food systems, and refrigerated transport systems.

• HFC-134a is acceptable for use in *new* equipment as a substitute for:

- CFC-11 and CFC-12 in centrifugal chillers.

- CFC-12 in reciprocating chillers, household refrigerators, cold storage warehouses, residential dehumidifiers, residential freezers, commercial ice machines, industrial process refrigeration, centrifugal chillers, refrigerated transport, retail food, vending machines, water coolers, and mobile air conditioners.

- CFC-500 in centrifugal chillers, dehumidifiers and refrigerated transport.

- CFC-502 in industrial process refrigeration and refrigerated transport.

• HFC-134a is acceptable for use in *existing* equipment, or retrofits, as an alternative for:

- CFC-12 in centrifugal chillers, reciprocating chillers, cold storage warehouses, residential dehumidifiers, industrial process refrigeration equipment, refrigerated transport systems, retail food systems, vending machines, and mobile air conditioners.

- CFC-500 in centrifugal chillers and refrigerated transport systems.

- CFC-502 in industrial process refrigeration equipment and refrigerated transport systems.

R-11 Alternative Refrigerants

The primary alternative refrigerant for the CFC R-11 is the HCFC R-123. A future option, which is not yet commercially available, is R-245 ca.

R-123

In the past, almost all centrifugal chillers used R-11. R-123 is the only commercially available alternative for converting these machines. The thermodynamic properties of R-123 are very similar to those of R-11.

R-123 is a high-efficiency, low-pressure refrigerant and is used only in centrifugal chillers. R-123 is safe and effective. Because of their molecule size, low-pressure refrigerants including R-11 and R-123 have lower density. This means chillers that use these refrigerants typically are larger than higher pressure machines. Because R-123 is competitively priced, has a very short atmospheric lifetime and very low ODP, GWP and cost, as well as nearly zero refrigerant loss, the development of other R-11 alternative refrigerants are a lower priority for manufacturers.[28]

R-123 is considered a "transitional alternative" or interim refrigerant. R-123 is an HCFC, it does contain chlorine, and a total production phase out of R-123 is scheduled by 2030. But during the next 35 years or so R-123 can be used, which is about the average lifetime of new equipment (approximately 30 years). A better long-term alternative would be a refrigerant without chlorine or bromine and zero (ODP).

Systems converted to use R-123 experience between three and 20% loss of efficiency. However, system changes can be made to improve efficiency. R-123 machines have a theoretical coefficients of performance (COP) of between 7 and 7.4. Actual performance chillers that use R-123 are commercially available with a COP of 6.76 (0.52 kW/ton).

R-123 is widely available and is being manufactured in large quantities. Both AlliedSignal and DuPont have plants that manufacture R-123.

R-123 has a safety classification of B1 and has an AEL (allowable exposure limit) of 30 ppm. With the required safety measures, R-123 is allowed to be used in the central plants of large commercial buildings. To answer some concerns about R-123 toxicity in the event of a leak, EPA states: *EPA believes that HCFC-123 is a necessary transition refrigerant as the world phases out the CFCs, and SNAP lists it as*

R-123 Summary	
Environmental	ODP: 0.02; HGWP 0.02
Interim or Long-term Alternative	Interim, contains chlorine, being phased out in 2030.
Safety	AEL (allowable exposure limit) of 30 ppm, Class B1
Relative Efficiency	COP 6.76 (52 kW/ton)
Availability	Widely available, produced by several manufacturers
Cost	From $4.00 to $7.00 per pound wholesale
Operating Pressure	Low-pressure refrigerant
Chemical behavior	Pure refrigerant (no glide)
Application	Low-pressure centrifugal chillers

Table 2-6 R-123 Summary

acceptable for use in chillers. It is safe to use in the long-term, and is actually safer in emergencies than CFC-11.[29]

R-245ca

A long-term alternative for R-11 or R-123 is R-245ca. This low-pressure refrigerant, which is chlorine-free, is still being tested and is not yet commercially available.

Preliminary tests of this new compound have shown it to have thermophysical properties very similar to those of R-11 and R-123. R-245ca has an ODP of zero and a global warming potential (GWP) about one-third that of R-134a. Toxicity testing of R-245ca has been very limited and conclusive results are not available; so far toxicity levels are low. Flammability may be a minor issue but if R-245ca is used in a blend, this concern could be resolved.

R-22 Alternative Refrigerants

R-22 is the predominant refrigerant in use today. It has a high operating efficiency and currently has a relatively low cost. R-22 is used in screw and reciprocating chiller systems as well as smaller "unitary" or package units.

Reciprocating chillers almost exclusively use R-22. This segment of the market, which accounts for almost one-third of the installed chiller capacity and two-thirds of the annual shipments of packaged chillers, is concerned with discovering appropriate alternative refrigerants.

Like R-123, R-22 is considered a "transitional alternative" or interim refrigerant. Depending on what's finally decided, R-22 is being phased out after year 2000 (the current EPA schedule calls for a phase-out of R-22 for new equipment applications by the year 2010, but it is possible it may be phased out for new applications as early as the year 2000). Since it is an interim refrigerant, the EPA does not specifically list substitutes for R-22. At this time, according to the EPA, any substitute is acceptable for R-22.

In this section, we'll discuss the three main replacement candidates for R-22:

- R-407C: zeotropic blend of HFCs (R-32/125/134a)

- R-410A: azeotropic mixture of R-32 and R-125

- R-134a: "pure" fluid

Other "pure" fluid alternatives for R-22 such as ammonia, and propane also are being considered, but there is not much interest in propane because it is flammable.

Of these candidates, the most promising is the R-32/125 azeotrope R-410A, because it has higher efficiency ratings.

Before we discuss the specific R-22 alternative refrigerants, we'll look at some issues regarding R-22 use in unitary equipment. The issues that are of interest regarding alternatives for R-22 as they relate to unitary systems crosses over to the search for alternatives for larger chiller systems that also use R-22.

Unitary Equipment and R-22

Unitary, or small package equipment consists of residential and light commercial air-conditioning systems. These small systems have an outdoor unit that houses the compressor and fan, and one or more indoor heat exchangers and blowers. R-22 is commonly used in these units, and the trend toward improving their efficiency continues.

The unitary equipment market is very competitive, and like other areas, is driven by consumer costs. Any alternative refrigerant that has considerably lower cooling capacities than R-22 are at a disadvantage when compared to alternatives with similar or higher cooling capacities. Any loss in capacity due to the performance of the refrigerant will need to be compensated for with larger compressors and heat exchangers, which cost more to produce and will end up costing the customer more.

Both "pure" refrigerants and blends are being explored as alternatives for R-22 in unitary equipment. Pure refrigerants such as propane (R-290), ammonia (R-717) and R-134a have been considered. Propane has similar thermophysical properties as R-22, but is flammable. (Propane is used widely in industrial refrigeration and has recently become the refrigerant of choice for small air conditioners for some European manufacturers. Ammonia has its own particular issues and is discussed later.) R-32 and R-152a have also been considered, but these are flammable and have not been pursued as pure refrigerants. (R-32 is found in a lot of blends. It has a low boiling point and low global warming potential number.)

Of the pure refrigerants considered, only R-134a has been considered as an alternative to R-22 in unitary air conditioning. It is a commercially available product (Carrier is one manufacturer). On the down side is the fact that R-134a has significantly lower capacity than R-22. R-134a has been impractical for unitary use because the volumetric capacity of an R-134a compressor must be about 50% larger than an R-22 compressor to achieve the same cooling capacity. Recent efforts toward replacing R-22 have gravitated toward refrigerant blends that have capacities similar to or better than R-22.

Theoretical tests were performed for alternative refrigerants for unitary equipment at the Oak Ridge National Laboratory, Oak Ridge, Tennessee, sponsored by the Alternative Fluorocarbons Environmental Acceptability Study (AFEAS), and the U.S. Department of Energy. The results for two test conditions provide some data on theoretical performance in terms of ideal steady-state COPs for selected HFC mixtures, ammonia, and propane.

Theoretical Steady-State COPs for Unitary Air Conditioners and Heat Pumps			
Refrigerant	Components	Operating Test Conditions	
		A*	B**
R-22	HCFC-22	100%	100%
R-134a	HFC-134a	100%	99%
R-717	Ammonia	102%	102%
R-290	Propane	99%	99%
R-410A	R-32/125 (50/50)	97%	98%
R-407C	R-32/125/134a (23/25/52)	101%	101%

* 35°C (95° F) ambient cooling, 10° C (50° F) evaporating, 46° C (115° F) condensing, 12.8° C (55° F) return gas, 40.6°C (105° F) liquid.
** 8.3°C (47° F) ambient heating, -1.1° C (30° F) evaporating, 33° C (92° F) condensing, 2° C (36° F) return gas, 33°C (92° F) liquid.

Table 2-7 Theoretical Steady-State COPs for Unitary Air Conditioners and Heat Pumps

The data from the tests show mixed results[30] (see the table above). The blends that show theoretical efficiencies equal or higher than R-22 have high temperature glides in the evaporator and condenser. R-134a and the zeotropic blends R-410A and R-407C have undergone compressor calorimeter tests and limited "drop-in" testing or testing in "soft-optimized" systems, but these tests were not performed in units optimized for these refrigerants. The potential efficiencies were not realized because the test units did not compensate for the temperature glides in the cross-flow air-to-air heat exchangers. Significant development efforts may realize the theoretically higher efficiencies in actual system operation.

Industry Efforts in Testing for Alternatives

The Air-Conditioning & Refrigeration Institute (ARI) has sponsored several programs created to evaluate alternatives for R-22 and R-502, including their Alternative Refrigerants Evaluation Program (AREP) and Materials Compatibility and Lubricants Research (MCLR) program.

The following are excerpts from a letter written by David S. Godwin, Engineer, Research Projects, Air-Conditioning & Refrigeration

Institute (ARI), Arlington, Virginia, dated March 20, 1995 that was sent to Mukesh K. Khattar, Manager, HVAC, Refrigeration and Thermal Storage, Electric Power Research Institute, Palo Alto, CA for presentation at Electric Power Research Institute (EPRI) workshop at the Southern California Edison Customer Technology Application Center (CTAC) in Irwindale, CA in 1995.[31]

In the letter, Mr. Godwin states:

"ARI's research efforts over the last few years have concentrated in the area of alternative refrigerants.

- **Materials Compatibility and Lubricants Research (MCLR) Program.** Over the past several years, we have managed research projects looking at the stability of new refrigerants and lubricants and their compatibility with plastics, elastomers [found in seals and gaskets] and motor materials.

- **Alternative Refrigerants Evaluation Program (AREP).** This program was established for industry to cooperate in the testing of numerous R-22 and R-502 alternatives. Organizations from the U.S., Canada, Europe and Japan participated in this program, which involved manufacturers testing their equipment with proposed alternatives.

- There seems to be growing interest to evaluate in refrigeration products some of the HFCs nominally offered as R-22 alternatives (e.g., R-410A&B and R-407C). Analysis of how to apply these refrigerants may help industry adjust to using these alternatives.

- R-407C is a zeotrope with significantly higher temperature glide than R-404A and R-507. If this zeotrope were applied in refrigeration products, some problems experienced in R-22 equipment could be intensified. For instance, maldistribution [sic] of air temperatures in heat exchangers led to premature coil frosting in some AREP tests."

AREP

As indicated in Mr. Godwin's letter, the Alternative Refrigerants Evaluation Program (AREP), operating under the coordination of the Air Conditioning and Refrigeration Institute (ARI), is an international cooperative program designed to identify alternative refrigerants for R-22 and R-502.

The AREP group, organized by ARI in 1991, currently consists of about 40 equipment, refrigerant, and compressor manufacturers from the United States, Canada, Europe, and Japan. Also participating are the U.S. Department of Energy, the National Institute of Standards and Technology, and Electric Power Research Institute (EPRI).

The AREP efforts are designed to provide data on the performance of air-conditioning and refrigeration equipment using ozone-friendly substances, and to take an active role in identifying replacement candidates for R-22 and R-502.

AREP established the testing and performance evaluation methodologies, and coordinates the refrigerant tests, and has impartially published the results. This group completed its screening program in 1994 and work continues. All of the members share in the selection, testing, and evaluation of candidate refrigerants. Also, the design of optimized equipment is left to the individual manufacturers. AREP will publish only the results and will not select final refrigerant alternatives.[32]

AREP Tests

AREP testing procedures attempt a fair evaluation of each refrigerant. The tests also provide information that can be used in determining actual system design.

There are four phases of AREP tests for each candidate refrigerant:

• First, the performance of each alternative is evaluated in compressors using calorimeter testing. These tests determine how well a given compressor operates with a particular refrigerant.

• Second, the refrigerant is tested in existing systems in "drop-in" tests. In these tests there is no modification to the equipment.

The calorimeter tests and "drop-in" tests have been completed and the data is currently available.

• Third, heat transfer tests are performed for the refrigerant under various conditions during both condensing and evaporating stages. These tests measure refrigerant-side heat transfer coefficients in "enhanced tubes."

• The final phase uses the data from the first three tests in system computer simulations and applies the information to "soft-optimized" systems where some system modification is implemented, or soft-optimized systems are designed, built, and tested.

AREP Testing Results

Final results of many AREP tests are available to the public through the ARI database. (You can contact ARI at (703) 524-8800.) Other data will soon be available.

So far, the information that's been accumulated confirms that there are several reasonable long-term alternatives for R-22 and R-502. While many do not demonstrate "ideal" characteristics, the results have shown actual system performance improvements may be possible using some of the newer refrigerants.

The AREP results show that many alternative refrigerants, when used in non-optimized compressors and systems, "can provide capacities and efficiencies close to, and sometimes exceeding, those achieved with the baseline refrigerant."[33]

So far, no one candidate refrigerant stands out as the obvious choice. The results indicate three possible directions regarding alternative refrigerants for R-22:

- Switch to a low-pressure refrigerant such as R-134a. Because R-134a systems operate at lower pressures than current R-22 systems, one drawback is that R-134a equipment would require much larger compressors and larger heat exchanger tubing.

- Choose from the zeotrope refrigerant alternatives such as R-407C. These blends have similar performance characteristics as R-22 and can accommodate traditional applications, but they do exhibit up to 10 degrees temperature glide, which is a concern for many people.

- Select from the high-pressure, higher-capacity refrigerant options such as R-410A or R-410B. Under most conditions, these operate at up to 50% higher pressures than R-22, which means they would require smaller compressor displacements and smaller tubing in the heat exchangers. Also, the heat exchangers would also need to be modified to accommodate the higher heat transfer rates of these refrigerants.

The rest of this section discusses R-407C, R-410A, and R-134a, the three main R-22 alternatives suggested from the AREP results.

R-407C

R-407C is a blend of R-32/125/134a (23/25/52). This zeotropic blend is a non-ozone-depleting refrigerant used as an alternative for R-22 and for R-12. (R-32 is found in a lot of blends. It has a low boiling point, and low GWP, but is flammable.)

R-407C has pressure, capacity, and efficiency characteristics that are very similar to R-22. This makes it a likely candidate for use in converted equipment or, in new equipment, as an alternative to R-22 in commercial and residential air conditioners and heat pumps. Because of its similarity to R-22, this refrigerant won't require significant changes in system design.

As a zeotrope, R-407C does exhibit glide characteristics. R-407C has on average a 12° F glide (boiling range). Preliminary reports indicate R-407C is comparable to R-22 in terms of system operation, but additional testing will be required to completely assess the performance of this refrigerant in different systems under varying operating conditions.

AREP tests (at evaporator temperatures of 45° F and condenser temperatures of 130° F) show that R-407C in existing R-22 systems increases capacity about 10% but reduces efficiency about 5%. The design of heat exchangers is an issue for systems that will use R-407C since R-407C, as a zeotropic blend, tends to be less efficient in transferring heat than pure, single-component refrigerants. This means heat exchangers will probably need to be larger than those currently used with R-22 systems and refrigerant-side heat exchanger tubes designed for R-22 will also need to be changed. These changes will mean higher costs. (See Chapter 3 for additional discussion on heat exchangers.)

The components of R-407C have undergone extensive toxicity testing by the Program for Alternative Fluorocarbon Toxicity Testing (PAFT). Results indicate the R-407C components have very low toxicity. The calculated Acceptable Exposure Limit (AEL), based on the AEL for each component, is 1000 ppm, 8- and 12-hour time-weighted average (TWA).

Table 2-8 summarizes the actual performance of R-407C when compared to R-22 in cooling and heating modes. Tests were performed using units of different sizes designed for R-22. The test units were not modified or optimized for performance with R-407C.

Performance of R-407C Relative to R-22 in Multiple Air Conditioners and Heat Pumps		
	R-407C Range of Performance	
	In Cooling Mode*	**In Heating Mode****
Relative Capacity	98% to 103%	93% to 106%
Relative Energy Efficiency Ratio (EER)	93% to 97%	94% to 97%
Change in Discharge Temperature	-15°F to -8°F (-8.3°C to -4.4°C)	-18°F to 0°F (-10°C to 0°C)
Change in Discharge Pressure	+15 psi to +40 psi	+9 psi to +34 psi

*   Values compared to R-22 in unmodified split system heat pumps and an unmodified window air conditioner using the DOE cooling test conditions A and B.

** Values compared to R-22 in unmodified split system heat pumps and an unmodified window air conditioner using the DOE heating test conditions E and H.

DuPont

Table 2-8   Performance of R-407C Relative to R-22 in Multiple Air Conditioners and Heat Pumps

R-407C Summary	
Environmental	ODP: zero; HGWP .28
Interim or Long-term Alternative	A long-term alternative blend, R-32/125/134a (23/25/52)
Safety	1000 ppm, 8- and 12-hour time weighted average (TWA)
Availability	Available through major manufacturers
Cost	Approximately $9.00 per pound
Chemical behavior	Zeotrope with from 11 to 13° F glide
Application	Air conditioning or refrigeration

Table 2-9   R-407C Summary

R-410A

R-410A is a blend of R-32 and R-125 (50/50). R-410A is an available substitute for R-22 in chillers and as a substitute for R-502 in commercial refrigeration. (R-410A is also an alternative for R-12 and R-502.) R-410B is another similar high-pressure blend, also composed of different percentages of R-32 and R-125.

When compared to R-22, R-410A operates at approximately 50% higher pressures at a given temperature (the compressor discharge pressure is approximately 180 psi higher). The higher pressures and high volumetric capacity means any system using R-410A will require thicker tubes and other structures to withstand the greater pressures, and compressors will need stronger bearings. But, since R-410A has a higher energy density than R-22, (a higher cooling effect) systems that use this blend will require compressors with smaller volume and smaller diameter tubing. Currently, there are no machines manufactured that use R-410A or R-410B.

R-410A is technically classified a zeotrope, but is in fact a near-azeotrope blend with very low temperature glide (0.2° F). (R-410A does exhibit a small composition shift—the liquid- and vapor-phase compositions are slightly different in air-conditioning systems.) So in spite of the higher operating pressures, R-410A can use traditional heat exchanger designs and will not have the fractionation problems associated with other blends with higher glide properties.

Because of the significant differences in operating pressures, R-410A only makes sense for use in new equipment built to accommodate the higher pressures (with compressor designs with stronger bearings and pressure-containing structures) and is not a suitable refrigerant for conversions.

With negligible glide, existing applications are possible if the system is changed to accommodate the higher pressures required by the refrigerant. This will simplify future system design, testing, and servicing.

AREP tests performed by AlliedSignal, using both scroll and reciprocating compressors, show that using R-410A in existing R-22 systems results in 40-50% more capacity. Also, their tests show that using this mixture in existing R-22 systems results in 5 to 13% lower efficiency at evaporator/condenser temperatures of 45°F/110°F. R-410A achieved a 5% efficiency gain over the baseline R-22, an approximate 10% gain

R-410A Summary	
Environmental	ODP 0; HGWP .44
Interim or Long-term Alternative	Long-term blend R-32/125 (50/50)
Safety	Rated A1
Refrigeration Capacity and Relative Efficiency	Preliminary tests show 40-50% more capacity than R-22; from 5-13% lower efficiency to 5% better efficiency compared to R-22
Availability	Available from AlliedSignal
Cost	Approximately $10.00 per pound
Operating Pressure	High discharge pressure and high volumetric capacity
Chemical Behavior	Low-glide zeotrope; near-azeotropic
Application	Air conditioning and refrigeration

Table 2-10 R-410A Summary

over R-407C, and an even greater efficiency gain over R-134a. Testing has not yet been done in an optimized system specifically designed for R-410A.[34]

R-134a

R-134a is a "pure" refrigerant that has zero ODP, and is an acceptable replacement for R-11 and R-12 in centrifugal chillers, for R-12 in reciprocating chillers, for R-500 and for R-502. Although R-134a has only been used on a limited basis in unitary or small package systems, it is a widely used substitute for R-22 in large chillers, as well as in automotive air conditioning and residential refrigerators.

As mentioned earlier, because the capacity of an R-134a compressor must be about 50% larger than an R-22 compressor to achieve the same cooling capacity, it is not currently a practical option for unitary equipment. AREP tests have shown that R-134a has between 30 to 40% less capacity than comparable units that use R-22. Tests at evaporator temperatures of 45° F and condenser temperatures of 130° F have shown efficiency variances of between -5% to +5% when compared to R-22.

R-134a Summary	
Environmental	ODP: 0; HGWP .28
Interim or Long-term Alternative	Long-term alternative
Safety	A1
Refrigeration Capacity and Relative Efficiency	Similar capacity and efficiency as R-12
Availability	Widely available
Cost	Because of a two-step manufacturing process, it has higher costs that HCFCs. $3.75 to $4.50/ per pound
Operating Pressure	Similar operating pressures as R-12
Chemical Behavior	Pure refrigerant with no glide
Application	Air conditioning and refrigeration; widely used in automobile air conditioners

Table 2-11 R-134a Summary

However, new designs using scroll compressors, instead of reciprocating compressors, may offset this loss of capacity and efficiency with more cost-effective designs. Carrier introduced unitary R-134a air conditioners in 1994 that use scroll compressors. These units have equal or better efficiency when compared to comparable R-22 units and cost only slightly more.

Summary Data for R-410A, R-407C and R-134a

The Electric Power Research Institute's Commercial Building Air-Conditioning Center provided summary data for R-22 alternatives R-410A, R-407C, and R-134a. The information provided in the following table gives a good overview comparison of these three candidates.[35]

Comparing R-22 Alternatives R-410A, R-407C, and R-134a			
	R-410A	R-407C	R-134a
Refrigerant Details			
Components	R-32/125	R-32/125/134a	R-134a
Weight %[a]	60/40	30/10/60	100%
Glide	0.2°F	8-10°F	none
Trade name	Genetron®AZ-20	Suva®AC9000/K LEA®66	several
Manufacturer	Allied Signal	DuPont/ICI	several
R-22 Comparison[b]			
Discharge pressure[c]	479	310	199
Volumetric capacity[d]	+40 to +50%	+10%	-30 to-40%
COP[e]	-5 to -13%	-5%	-5 to +5%
Application Tradeoffs			
Impact of use with minimal system change	not applicable	similar efficiency and capacity	much lower capacity
System changes need to optimize performance	compressor designed for higher pressure & lower refrig. flow	new counter flow heat exchangers	larger compressor

[a] Composition tested by AREP members. Current composition is slightly different.

[b] Approximate R-22 comparisons. AREP compressor calorimeter data from a paper by David Godwin, of ARI, presented at August 1993 ASHRAE/NIST Refrigerants Conference.

[c] Pressure measured as psig at 130°F condensing; R-22 is 297 psig.

[d] Tons/cfm.

[e] Coefficient of performance. Initial compressor comparison only, at 45°F evaporating/110-130°F condensing.

Table 2-12 Comparing R-22 Alternatives R-410A, R-407C, and R-134a

R-12 and R-502 Alternatives

There are several alternative refrigerants that are potential substitutes for R-12 and R-502, two CFCs that are being phased out. They are mostly blends, but some pure refrigerant alternatives are also considered.

Some of the alternatives discussed below include some of the test results of R-502 alternatives performed under AREP. (For more information about AREP tests and how to obtain the results, contact ARI.)[36]

R-507

R-507 is a blend of R-125/143a (45/55) and is classified an azeotrope.

AREP-sponsored testing of R-507 compared it to R-502 in reciprocating and scroll compressors and also tested to compare R-507 to R-22.

R-507 in systems with a reciprocating compressor and a scroll compressor and at low condensing temperatures has better capacity than R-502, a 14% improvement in a reciprocating system and 9% in a scroll system.

R-507 was tested and compared to R-22 with varying results regarding capacity. The tests showed that at higher condensing temperatures R-507 has higher capacity as compared to R-22, but has lower capacity at lower evaporating temperatures. Capacity for R-507 when compared to R-22 in scroll and rotary compressors is somewhat better.

When compared to R-502, R-507 has 5% better efficiency in reciprocating compressors and 6% better efficiency in scroll compressors. Efficiency ratios of the scroll compressor was 0.89 and for the rotary compressor was 0.98.

R-507 Summary	
Environmental	ODP 0; HGWP .98
Interim or Long-term Alternative	Long-term alternative; R-125/143a (45/55)
Safety	A1
Refrigeration Capacity	Low-temperature, R-502 replacement
Availability	Available from AlliedSignal
Cost	$7.25 to $7.50 per pound
Operating Pressure	High-pressure
Chemical Behavior	No glide
Application	Refrigeration

Table 2-13   R-507 Summary

R-404A

R-404A is a zeotrope blend of R-125/143a/134a (44/52/4).

AREP data indicates that when compared to R-502, R-404A capacity decreases with increasing condensing temperatures. At lower condensing temperatures, R-404A has about the same capacity as R-502, but rapidly decreasing capacity at higher condensing temperatures.

For a fixed evaporating temperature, R-404A has lower efficiency when compared to R-502 when condensing temperatures increase. R-404A efficiency is less when compared to R-502 at high condensing temperatures and better for low temperatures. At 90° F to 105°F condensing temperatures, the efficiency levels are the same for the two refrigerants.

R-404A Summary	
Environmental	ODP 0; HGWP .94
Interim or Long-term Alternative	Long-term alternative; R-125/143a/134a (44/52/4)
Safety	A1
Refrigeration Capacity	About the same as R-502 at lower condensing temperatures, decreasing capacity at higher condensing temperatures
Availability	Available from DuPont, AlliedSignal, Elf Atochem
Cost	$6.70 to $8.50 per pound
Operating Pressure	High-pressure
Chemical Behavior	Minimal (1° F) glide
Application	Refrigeration

Table 2-14 R-404A Summary

R-407A and R-407B

R-407A and R-407B are long-term HFC replacement refrigerants for R-502 for low-temperature applications. R-407A and R-407B are three-part zeotropic blends of R-32, R-125, and R-134a that have been independently tested by Bitzer, Searle, and other manufacturers. Their performance is similar to R-502 in a wide range of low-temperature applications.

With low GWP and zero ODP, these blends have minimum environmental impact. They have good energy efficiency, are non-flammable, and are immediately available.

R-407A is appropriate for new system designs that can take advantage of the energy efficiency potential of these blends, and for converted systems that have moderate condensing temperatures and/or medium to high evaporator temperatures.

R-407B is appropriate for converting existing R-502 equipment in applications where compressor discharge temperature must be maintained at R-502 levels.

The flowchart below from ICI gives some guidelines for choosing between R-407A and R-407B. For conversions with high lift (-40° F evaporator with 105° F or higher condenser condition) use R-407B. For conversions with moderate lift (-20° F evaporator with 120° F or higher

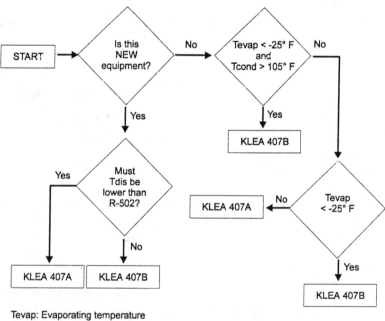

Tevap: Evaporating temperature
Tcond: Condensing temperature
Tdis: Discharge temperature

Figure 2-13 Choosing Between R-407A and R-407B

R-407A Summary	
Environmental	ODP 0; HGWP .49
Interim or Long-term Alternative	Long-term HFC alternative; R-32/125/134a (20/40/40)
Safety	A1; non-flammable
Availability	Widely available; ICI produces KLEA 60
Cost	Approximately $6.75 per pound
Operating Pressure	High-pressure
Chemical Behavior	13° F glide
Application	Refrigeration

Table 2-15   R-407A Summary

R-407B Summary	
Environmental	ODP 0; HGWP .70
Interim or Long-term Alternative	Long-term HFC alternative; R-32/125/134a (10/70/20)
Safety	A1; non-flammable
Availability	Widely available; ICI produces KLEA 61
Cost	Approximately $6.75 per pound
Operating Pressure	High-pressure
Chemical Behavior	5° F glide
Application	Refrigeration

Table 2-16   R-407B Summary

condenser condition) use either R-407A or R-407B. For conversions with low lift (-15° F evaporator with 105° F or higher condenser condition) use R-407A.

R-410A

R-410A is also an alternative for R-12 and R-502 as well as an alternative for R-22. R-410A is an azeotropic mixture of R-32 and R-125. This refrigerant is discussed earlier.

R-152a

R-152a to date is not used as a separate refrigerant, and is only used in blends. It has a very low GWP and has similar compatibility as R-134a.

Potential problems with R-152a are that it is flammable and it has low thermal stability and cannot be used in systems with a high discharge temperature. There is some discussion of blending R-152a with R-134a to reduce its flammability to non-critical levels. However, it requires a large amount of R-134a to make the mixture non-flammable.

R-125

R-125 is primarily used in blends, but also is used as a "secondary" loop refrigerant. It is non-flammable, and operates at low discharge temperatures. Because it has a low critical temperature, it cannot be used effectively with air-cooled equipment. R-125 takes good advantage of liquid subcooling and may be usable in special applications in blends. R-125 has an HGWP of .84.

R-22 as an Option for R-12 and R-502 in Refrigeration

R-22 is a "pure" refrigerant that in certain situations is an appropriate interim alternative for R-12 and R-502. R-22 was essentially the only option available to the commercial refrigeration industry when the CFC phaseout began. Most users converted new low-temperature system designs from R-502 to R-22. Many medium-temperature systems were already using R-22 or if not, changed over quickly.

Even as R-134a became available as a medium-temperature option, many applications such as supermarkets stayed with R-22 due to the larger physical compressor and piping sizes associated with R-134a. Also, R-22 has good efficiency characteristics on medium-temperature applications. New HFCs designed to replace R-502 are not as efficient as R-22 in medium-temperature applications.

R-22 continues to be used extensively in commercial and air conditioning applications. As an HCFC scheduled for eventual phaseout, why is R-22 continued to be used so much? Some reasons are that R-22 is relatively inexpensive, has a relatively low HGWP (.28), and it has known characteristics, compressor performance, and oil compatibility.

R-22 is inexpensive because it's easy to manufacture, there is existing production capacity, and it isn't taxed. (If policy regulators see a tax on HCFCs as an incentive to change to long-term alternatives and as a potential for revenue, R-22 may be taxed in the future.)

Supply of R-22 will be adequate even with production caps (R-22 production levels will gradually decrease: they are limited to 1986 levels in 1995, and by year 2003, must be limited to 65% of 1986 production levels. The production cap on R-22 in 1996 is approximately 220 million pounds). R-22 production in 1994 was reported as 307 million pounds, a 5% increase over 1993.

Many manufacturers have invested in R-22. On the other hand, availability will depend somewhat on how much CFCs are reclaimed. Most reclaimed refrigerant is R-22. R-22 use is increasing both in new construction and as a replacement for CFCs directly and as a component of the interim blends.

R-22-based Interim Blends as Options for R-12 and R-502

A number of interim blends that are alternatives for R-12 and R-502 refrigeration systems have become available, and most contain from 40 to 60% R-22. Because they contain R-22, they are considered interim blends since R-22 eventually will be phased out.

R-22 Summary	
Environmental	ODP .055; HGWP .35
Interim or Long-term Alternative	Interim alternative; phaseout starts in 2004, total phaseout in 2030
Safety	A1 *Non-Flammable*
Refrigeration Capacity and Relative Efficiency	Capacity compatible with air conditioning and medium-temperature refrigeration; high efficiency
Availability	Widely available
Cost	$1.50 to $2.00 per pound
Operating Pressure	High-pressure
Chemical Behavior	Pure chemical, no glide
Application	Air conditioning and medium-temperature refrigeration

Table 2-17 R-22 Summary

A. Compounds have a relatively low toxicity rate.

A1. No flame propagation — #1 Compounds are not flamable. Low chronic Toxicity

Interim Blend Refrigerant	Typical Application
R-401A, R-401B	R-12 retrofit
R-402A, R-402B	R-502 retrofit
R-403	R-502 retrofit
R-405A	R-12 retrofit
R-406A	R-12 retrofit
R-408A	R-502 retrofit
R-409A	R-12 retrofit
R-411A, R-411B	R-22 service

Table 2-18 Interim Blends and Typical Applications

New refrigerant blends are appearing on the market with surprising frequency, and as the phaseout continues, the number of these customized blends will probably increase. The major refrigerant manufacturers offer products that have passed the scrutiny of the industry at large. However, products from less well-known manufacturers should be closely scrutinized before they are used—pay careful attention to such issues as flammability, toxicity, and reliability, refrigerant manufacturer equipment warranties, and future recovery and recycling capabilities.

For more information on these refrigerants, see the Commercial Refrigerants Database in this chapter and "Interim Refrigerant Choices" in Chapter 4.

R-290 (Propane)

Propane (R-290) is an R-12 and R-502 alternative that has been used worldwide in industrial plants with proven results. It is an interesting alternative in that despite its flammability, it has zero ODP and zero direct global warming effect.

R-290 can be used with copper tubing and hermetic compressors and has pressures and efficiency comparable to R-22.

One heat pump manufacturer in Germany has converted its residential heat pump line to propane, and propane can also be used in domestic refrigerators.

R-290 (Propane) Summary	
Environmental	ODP 0; HGWP 0
Interim or Long-term Alternative	Long-term alternative
Safety	A3, flammable
Refrigeration Capacity and Relative Efficiency	Good
Availability	Widely available
Cost	High-pressure (similar to R-12)
Operating Pressure	Pure substance, no glide
Chemical Behavior	Refrigeration
Application	Air conditioning and medium-temperature refrigeration

Table 2-19 R-290 (Propane) Summary

R-717 (Ammonia)

Ammonia has a number of very attractive characteristics, including zero ozone depletion, zero direct global warming effect, low cost, high theoretical efficiency, and a low required mass flow per ton of refrigeration.

Ammonia (NH_3) (R-717) has been used as a refrigerant for well over 100 years and is without a doubt the best refrigerant in many ways. It has more than six times the "latent heat capacity" of R-22, and requires a much smaller refrigerant charge for a given cooling load. In addition, its superior ability to transfer heat means R-717 machinery requires smaller evaporators and condensers.

Today's screw compressors with a single-stage compressor can tolerate the high discharge temperature of NH_3. Ammonia costs only 10 to 20% the cost of halocarbon refrigerants. This is why ammonia is the predominant refrigerant used in the food production industry—everything from fresh produce cooling and storage, to dairies, breweries, and wineries.

R-717 isn't necessarily a good centrifugal chiller refrigerant because it would need many stages of compression to get the required temperature lift; however, screw compressors and chillers are available in nearly the same size as centrifugal chillers and can achieve the required lift.

So why is ammonia not used more in refrigeration? Ammonia can cause stinging eyes, burning nose, and difficulty breathing which means ammonia must be carefully contained. However, if an accidental release occurs, today's improved techniques for detection and harmless dispersion can usually avoid serious injury.

There are various ratings for allowed exposure limits (AEL) for NH_3. An acceptable level might be 25 ppm, which is about equal to the level detected after mopping a kitchen floor. If monitoring can be as sensitive as 1 ppm, as it is with halocarbons, then there shouldn't be too much of a problem. R-123, a centrifugal chiller refrigerant, has an AEL of 30 ppm and with the required safety measures, is allowed to be used in the central plants of large commercial buildings. Should there be an accidental release of R-123, those people in and around the building would probably never detect it. If the same problem occurred with R-717, there probably would be panic. This is the risk that most commercial property owners are not willing to take.

A remote cooling plant some distance from an occupied facility is one way to possibly mitigate potential problems from accidental release. A central but isolated heating and cooling plant could supply chilled and hot water for millions of square feet of buildings.

Ammonia could prove to be a distant alternative for R-22; at one time R-12, R-502 and R-22 were considered alternatives to ammonia. Because ammonia has zero ozone depletion and zero global warming potential, and in spite of the increased energy use and higher TEWI associated with ammonia, it may end up as the choice over the "new" HFCs in indirect refrigeration applications.

R-717 (Ammonia) Summary	
Environmental	ODP 0; HGWP 0
Interim or Long-term Alternative	Long-term alternative
Safety	B2, toxic and flammable
Refrigeration Capacity and Relative Efficiency	Good
Availability	Widely available
Cost	$0.25 to $0.50 per pound
Operating Pressure	High-pressure
Chemical Behavior	Pure substance, no glide
Application	Air conditioning and refrigeration

Table 2-20   R-717 (Ammonia) Summary

FICs (Fluoroiodocarbons)

Other distant alternatives for CFCs is a new group of chemicals known as fluoroiodocarbons (FICs). These compounds contain fluorine, iodine, and carbon in various molecular configurations.

One FIC (CF_3I) blended with R-152a (51/49%) was recently used in an R-12 refrigerator without an oil change. The refrigerator has "accumulated over 1,500 hours of operation without apparent ill effects."[37] However, these chemicals are not particularly stable and iodine can be highly corrosive.

Other Alternative Refrigerant Issues

There are three additional issues that are important to consider when retrofitting or converting chillers and commercial refrigeration systems to use an alternative refrigerant:

- Conversion oil changing.
- Saturation pressures.
- Tube design.

Conversion Oil Changes

In the course of converting machinery to an alternative refrigerant, it is important to note that most of the new low-ozone-depleting refrigerants do not work well with the mineral oils that are currently used. HFC refrigerants will not work at all with mineral oil. This means new lubricants compatible with the new refrigerant are necessary when implementing an alternative refrigerant.

In addition, when changing oils in conversions, a series of several oil changes are required before a converted system is ready to accommodate a new refrigerant.

This is not an issue for new systems. A compatible oil is provided with the system.

(See Chapter 4 for additional discussion on refrigeration system lubricants.)

High Vapor Pressures and Tube Design

Alternative refrigerants may operate at higher pressures and may require changing refrigerant tubing. Operational safety becomes an important issue especially if the original tube wall thickness was not designed to handle high pressures.

Tubing used in any system must be properly rated, especially in systems that must handle the possible high pressures experienced with some of the new refrigerants. This is essential if hot gas defrost is used for an evaporator coil and exposes high pressure to the low pressure system side where tubing is usually larger in diameter. Larger diameter tubing is at greater risk of not meeting the required pressure rating. (For additional information, there are graphs under "System Design Pressures" in Chapter 4 that show the saturation pressures at 130° F for various refrigerants and the required tube design.)

COMMERCIAL REFRIGERANTS DATABASE

The series of tables that follow list traditional refrigerants currently in use, new refrigerants in use, and some new refrigerants under consideration.

The tables include a variety of information including typical uses, ODP, GWP, whether it is flammable or not, the ASHRAE Class and Exposure rating, and other information.

Common Refrigerants—Sorted by ASHRAE Number

ASHRAE Number	Composition Prefix	Chemical Formula/Composition	ODP	HGWP	Notes/Uses
11	CFC	CCl_3F (methane series)	1.0	1.00	Most centrifugal compressors, chillers; End of Production (EOP) 1995
12	CFC	CCl_2F_2 (methane series)	1.0	2.09	Reciprocating compressors, some centrifugals; EOP 1995
22	HCFC	$CHClF_2$ (methane series)	0.055	0.43	Phaseout starts 1996; EOP 2030
32		CH_2F_2 (methane series)	0.0	0.14	Mildly flammable
123	HCFC	$CHCl_2CF_3$ (ethane series)	0.02	0.02	Low-pressure centrifugals; phase out begins in 1996, EOP 2030
125	HFC	CHF_2CF_3 (ethane series)	0.0	0.84	Low critical temperature, which may limit use as a pure fluid
134a	HFC	CH_2FCF_3 (ethane series)	0.0	0.28	First HFC; autos, chillers; new or retrofit R-12 compressors
245ca	HFC	$CH_2F\text{-}CF_2\text{-}CHF_2$ (propane series)	0.0		Possible future R-123 replacement
290	HC	$CH_3CH_2CH_3$ (propane)	0.0	0.0	Small refrigerators or industrial equipment
401A	HCFC/HFC	R-22/152a/124 (53/13/34)	0.036	0.22	Interim alternative above -10°F; transition for R-12
401B	HCFC/HFC	R-22/152a/124 (61/11/28)	0.038	0.24	Interim alternative below -10°F; transition for R-12
402A	HCFC/HC/HFC	R-125/290/22 (60/2/38)	0.022	0.63	Interim alternative; transition for R-502
402B	HCFC/HC/HFC	R-125/290/22 (38/2/60)	0.033	0.52	Interim alternative; transition for R-502
403A	HCFC/HFC/HC	R-290/22/218 (5/75/20)	0.041	1.19	Interim alternative; commercial, transport refrigeration; transition for R-502
403B	HCFC/HFC/HC	R-290/22/218 (5/56/39)	0.033	4.09	Interim alternative; transition for R-502
404A	HFC	R-125/143a/134a (44/52/4)	0.0	0.94	Long-term R-502 and R-22 alternative; new installations
405A	HCFC/HC	R-22/152a/142b/C318 (45/7/5.5/42.5)	0.024	0.22	Interim option; transition refrigerant for R-12
406A	HCFC/HC	R-22/600a/142b (55/4/41)	0.060	0.50	Mobile air-conditioning; R-12 alternative

Table 2-21 Common Refrigerants—Sorted by ASHRAE Number

Common Refrigerants—Sorted by ASHRAE Number (continued)

ASHRAE Number	Composition Prefix	Chemical Formula/Composition	ODP	HGWP	Notes/Uses
407A	HFC	R-32/125/134a (20/40/40)	0.0	0.49	Long-term replacement for R-502; retrofit and new applications
407B	HFC	R-32/125/134a (10/70/20)	0.0	0.7	Close to existing R-502 conditions; retrofit and new applications
407C	HFC	R-32/125/134a (23/25/52)	0.0	0.28	Close to existing R-22 conditions
408A	HCFC/HFC	R-125/143a/22 (7/46/47)	0.024	0.75	Interim retrofit solution for R-502 medium- and low-temperature refrigeration
409A	HCFC	R-22/124/142b (60/25/15)	0.05	0.3	Interim alternative; R-12 transition
410A	HFC	R-32/125 (50/50)	0.0	0.44	High pressure (50% >R-22); needs compressor redesign; replacement for R-22 and R-502
410B	HFC	R-32/125 (45/55)	0.0	0.4	High pressure (50% >R-22); needs compressor redesign; replacement for R-22 and R-502
411A	HCFC/HC	R-1270/22/152a (1.5/87.5/11)	0.035	0.33	Interim option; transition refrigerant for R-22
411B	HCFC/HC	R-1270/22/152a (3/94/3)	0.037	0.35	Interim option; transition refrigerant for R-502
412A	HCFC/HFC	R-22/218/142b (70/5/25)	0.052		Interim R-500 alternative; ultra-low-temperature
500	CFC/HFC	R-12/152a (73.8/26.2)	0.545		Centrifugal chillers, heat pump water heaters
502	CFC/HCFC	R-22/115 (48.8/51.2)	0.180	3.75	Used in reciprocating compressors in food service; EOP 1995
507	HFC	R-125/143a (45/55)	0.0	0.98	New or retrofit applications designed to replace R-502
508A	HFC	R-23/116 (39/61)	0.2	4.1	Very low-temperature applications
717	IC	NH_3·ammonia	0.0	0	Industrial compressors

Note: The information is derived from various sources. Verify accuracy with refrigerant manufacturers.

Table 2-21   Common Refrigerants—Sorted by ASHRAE Number (continued)

ASHRAE Number	Pure Compound or Blend	Boiling Point/ Range (approx.)	Evaporative Glide (approx.)	ODP	Saturation Pressures psig at -30°F	psig at 20°F	psig at 120°F	Safety Class** Toxicity Limit
11	Pure Compound	75° F	0° F	1.0	27.8*	21.1*	18.5	A1/1000
12	Pure Compound	-22° F	0° F	1.0	5.5*	21.1	157.3	A1/1000
22	Pure Compound	-41° F	0° F	0.055	4.9	43.1	260.0	A1/1000
32	Pure Compound	-61° F	0° F	0.0	18.2	80	429.5	A2/1000
123	Pure Compound	82° F	0° F	0.02	28.3*	22.8*	15.1	B1/30
125	Pure Compound	-55° F	0° F	0.0	12.8	63.8	344.6	A1/1000
134a	Pure Compound	-15° F	0° F	0.0	9.8*	18.4	171.1	A1/1000
245ca	Pure Compound	77° F	0° F	0.0				
290	Pure Compound	-44° F	0° F	0.0	5.7	41.2	228.4	A3/1000
401A	Zeotropic Blend	-27° to -16° F	9° F	0.036	9" vac	18.1	165.9	A1/A1/800
401B	Zeotropic Blend	-30° to -19° F	9° F	0.038	7" vac	21.5	180.0	A1/A1/840
402A	Zeotropic Blend	-56° to -53° F	3° F	0.022	11.3	59.3	325.2	A1/A/1000
402B	Zeotropic Blend	-53° to -50° F	3° F	0.033	8.9	53.5	302.8	A1/A1/1000
403A	Zeotropic Blend	-58° F	4° F	0.041				A1/A1
403B	Zeotropic Blend	-59° F	2° F	0.033				A1/A1
404A	Zeotropic Blend	-52° F	1° F	0.0	9.9	55.5	309.7	A1/A1/1000
405A	Zeotropic Blend	-17.1° F	high	0.024				A1/A1

Table 2-22 Refrigerant Properties—Sorted by ASHRAE Number

Refrigerant Properties—Sorted by ASHRAE Number

ASHRAE Number	Pure Compound or Blend	Boiling Point/ Range (approx.)	Evaporative Glide (approx.)	ODP	Saturation Pressures			Safety Class** Toxicity Limit
					psig at -30°F	psig at 20°F	psig at 120°F	
406A	Zeotropic Blend	-26° to ? F	high	0.06				A1/A2
407A	Zeotropic Blend	-49° to -38° F	10° F	0.0	3.5	42.8	285.3	A1/A1/1000
407B	Zeotropic Blend	-53° to -45° F	6° F	0.0	7.3	52.9	319	A1/A1/1000
407C	Zeotropic Blend	-49° to -36° F	11° F	0.0	1.5	36.9	261.7	A1/A1/1000
408A	Zeotropic Blend	-46° F	1° F	0.024	7.6	49.5	286.8	A1/A1/1000
409A	Zeotropic Blend	-30° to -15° F	12° F	0.05	9.9	18.5	171.1	A1/A1/1000
410A	Zeotropic Blend	-62.5° F	0° F	0.0	18	79	416	A1/A1/1000
410B	Zeotropic Blend	-61° F	0° F	0.0				A1/A1
411A	Zeotropic Blend	-39° F		0.42	3.6	39	243	A1/A2
411B	Zeotropic Blend	-41° F		0.45	4.9	43	255	A1/A2
412A	Zeotropic Blend	-37° F		0.52				A1/A2
500	Azeotropic Blend	-28° F	0° F	0.545				A1/1000
502	Azeotropic Blend	-50° F	0° F	0.23	9.2	52.5	282.72	A1/1000
507	Azeotropic Blend	-51° F	0° F	0.0	10.3	57.624	321.97	A1/1000
508A	Azeotropic Blend	-122° F	0° F	0.0				A1/1000
717	Inorganic Compound	-28° F	0° F	0.0	1.6*	33.5	271.9	B2/25

* Inches Hg Vacuum   ** (ppm) AEL or other

Note: The information is derived from various sources. Verify accuracy with refrigerant manufacturers.

Table 2-22 Refrigerant Properties—Sorted by ASHRAE Number (continued)

			Refrigerant Properties—Sorted by Composition Prefix		
Composition Prefix	ASHRAE Number	Manufacturer	Chemical Formula Composition	Common Trade Names	Lubricant Oil
CFC	11	Many	CCl_3F-methane series	Genetron 11, Freon 11, Forane 11, Arcton 11	Mineral Oil (MO)
CFC	12	Many	CCl_2F_2-methane series	Genetron 12, Freon 12, Forane 12, Arcton 12	MO
CFC/HCFC	502	Many	R-22/115 (48.8/51.2)	Genetron 502, Freon 502, Forane 502, Arcton 502	MO
CFC/HFC	500	AlliedSignal, DuPont	R-125/12/152a (73.6/26.2)	Genetron 500, Freon 500,	MO
HC	290		$CH_3CH_2CH_3$- propane	Propane	
HCFC	22	Many	$CHClF_2$-methane series	Genetron 22, Freon 22,Forane 22, Arcton 22	MO
HCFC	123	AlliedSignal, DuPont, Elf Atochem	$CHCl_2CF_3$-ethane series	Genetron-123, SUVA Centri-LP, Forane 123	MO
HCFC	409A	Elf Atochem	R-22/124/142b (60/25/15)	FX-56	MO and/or Alkylbenzene (AB)
HCFC/HC	405A	China Sun, Greencool	R-22/152a/142b/C318 (45/7/5.5/42.5)	China Sun G2015, GU G2015, Greencool 12	MO/AB
HCFC/HC	406A	Monroe Air Tech	R-22/600a/142b (55/4/41)	GHG12	MO+AB
HCFC/HC	411A	Greencool	R-1270/22/152a (1.5/87.5/11)	G2018A	MO/AB
HCFC/HC	411B	Greencool	R-1270/22/152a (3/94/3)	G2018B	MO/AB
HCFC/HC/HFC	402B	DuPont	R-125/290/22 (38/2/60)	SUVA HP-81	MO+AB
HCFC/HC/HFC	402A	DuPont, AlliedSignal	R-125/290/22 (60/2/38)	SUVA HP-80,Genetron HP80	MO+AB
HCFC/HFC	401B	DuPont, AlliedSignal	R-22/152a/124 (61/11/28)	SUVA MP-66, Genetron MP66	AB
HCFC/HFC	401A	AlliedSignal, DuPont	R-22/152a/124 (53/13/34)	Genetron MP-39, SUVA MP-39	AB
HCFC/HFC	408A	Elf Atochem	R-125/143a/22 (7/46/47)	FX-10	MO/AB/Polyol Ester (POE)
HCFC/HFC	412A	ICI, LaRoche	R-22/218/142b (70/5/25)	Arcton TP5R	

Table 2-23 Refrigerant Properties—Sorted by Composition Prefix

Refrigerant Properties—Sorted by Composition Prefix (continued)

Composition Prefix	ASHRAE Number	Manufacturer	Chemical Formula Composition	Common Trade Names	Lubricant Oil
HCFC/HFC/HC	403A	Rhone Poulenc, Star Refrigeration	R-290/22/218 (5/75/20)	Isceon 69S, Starton 69	MO+Alkylbenzene
HCFC/HFC/HC	403B	Rhone Poulenc	R-290/22/218 (5/56/39)	Isceon 69L	
HFC	32		CH_2F_2-methane series		
HFC	125		CHF_2CF_3-ethane series		
HFC	134a	Many	CH_2FCF_3-ethane series	Genetron 134a, SUVA Cold MP, Forane 134a, KLEA 134a	Polyol Ester (POE)
HFC	245ca		propane series		
HFC	404A	DuPont, AlliedSignal, Elf Atochem	R-125/143a/134a (44/52/4)	SUVA HP-62, Genetron 404A, Forane 404A, FX-70	POE
HFC	407A	ICI	R-32/125/134a (20/40/40)	KLEA 60	POE
HFC	407B	ICI	R-32/125/134a (10/70/20)	KLEA 61	POE
HFC	407C	AlliedSignal, ICI, DuPont	R-32/125/134a (23/25/52)	Genetron 407C,KLEA 66,SUVA 9000	POE
HFC	410A	AlliedSignal	R-32/125 (50/50)	AZ20	POE
HFC	410B	DuPont	R-32/125 (45/55)	SUVA 9100	POE
HFC	507	AlliedSignal	R-125/143a (45/55)	Genetron AZ-50	POE
HFC	508	ICI	R-23/116 (39/61)	KLEA 5R3	
IC	717	Many	NH_3 ammonia	Ammonia	

Note: The information is derived from various sources. Verify accuracy with refrigerant manufacturers.

Table 2-23　Refrigerant Properties—Sorted by Composition Prefix (continued)

New Alternative Refrigerants

Alternatives for...	ASHRAE Number	Manufacturers	Notes/Uses
R-12	134a	AlliedSignal, DuPont, Elf Atochem, ICI	First HFC, autos, chillers. New or retrofit R-12 compressors
	401A	AlliedSignal, DuPont	Interim alternative (contains HCFC-22) above -10° F
	401B	AlliedSignal, DuPont	Interim alternative (contains HCFC-22) below -10° F; transition for R-12
	405A	China Sun, Greencool	Interim alternative (contains HCFC-22); transition refrigerant
	406A	Indianapolis, Monroe Air Tech	Mobile air-conditioning; R-12 alternative
	409A	AlliedSignal, Elf Atochem	Interim alternative (contains HCFC-22), transition for R-12
R-22	407C	AlliedSignal, DuPont, Elf Atochem, ICI	Close to existing R-22 conditions
	410A	AlliedSignal	Compressor redesign; >50% higher pressure than R-22; new installations long-term R-22 or R-502 alternative
	410B	DuPont	Compressor redesign; >50% higher pressure than R-22; new installations long-term R-22 or R-502 alternative
	411A	Greencool	Interim R-22
R-502	402A	AlliedSignal, DuPont	Interim alternative; transition for R-502
	402B	AlliedSignal, DuPont	Interim alternative; transition for R-502
	403A	Rhone Poulenc	Interim alternative (contains HCFC-22); transition for R-502
	404A	AlliedSignal, DuPont, Elf Atochem	Long-term R-502 and R-22 alternative; new installations
	407A	ICI	Long-term replacement for R-502; retrofit and new applications
	407B	ICI	Close to existing R-502 conditions; retrofit and new applications
	408A	Elf Atochem	Interim retrofit solution for medium and low temperature refrigeration systems using R-502
	411B	Greencool	Interim alternative; transition for R-502
	507	AlliedSignal, Elf Atochem	New or retrofit applications designed to replace R-502
R-11	123	AlliedSignal, DuPont, Elf Atochem	Low-pressure centrifugals; phased out in 1996

Table 2-24 New Alternative Refrigerants

Refrigerant Alternatives—Sorted by Manufacturer

ASHRAE #	Notes and Uses
AlliedSignal (Genetron, AZ-xx)	
11	Most chillers with centrifugal compressors; End of Production (EOP) 1995
12	Reciprocating compressors, some centrifugal compressors; EOP 1995
22	Reciprocating, screw, & scroll compressors; phaseout starts 1996, EOP 2030
123	Low-pressure centrifugals; phased out in 1996
134a	First HFC, autos, chillers; new or retrofit R-12 compressors
401A	Interim alternative above -10° F (contains HCFC-22)
401B	Interim alternative below -10° F (contains HCFC-22)
402A	Interim alternative (contains HCFC-22); transition for 502
404A	Long-term R-502 and R-22 alternative; new installations
407C	Close to existing R-22 conditions
409A	Interim alternative, (contains HCFC-22), R-12 transition refrigerant
410A	>50% higher pressure than R-22; requires compressor redesign; long-term replacement for R-22 or R-502.
500	Centrifugal chillers, heat pump water heaters
502	Reciprocating compressors in food service; EOP 1995
507	New or retrofit applications designed to replace R-502
DuPont (Freon, MP-xx, SUVA HP-xx, SUVA AC-xxxx)	
11	Most centrifugal compressors chillers; (EOP) 1995
12	Reciprocating compressors, some centrifugal compressors; EOP 1995
123	Low-pressure centrifugals; phased out in 1995
134a	First HFC, autos, domestic, chillers. New or retrofit CFC-12 compressors
401A	Interim alternative above -10° F (contains HCFC-22)
401B	Interim alternative below -10° F(contains HCFC-22)
402A	Interim alternative (contains HCFC-22); transition for R-502
402B	Interim alternative (contains HCFC-22) ; transition for R-502
404A	Long-term R-502 and R-22 alternative; new installations
407C	Close to existing R-22 conditions
410B	>50% higher pressure than R-22; requires compressor redesign; long-term replacement for R-22 or R-502.

Table 2-25   Refrigerant Alternatives—Sorted by Manufacturer

Refrigerant Alternatives—Sorted by Manufacturer (continued)

ASHRAE #	Notes and Uses
Elf Atochem (Forane FX-xx, Forane-xx)	
11	Most centrifugal compressors chillers; End of Production (EOP) 1995
12	Reciprocating compressors, some centrifugal compressors; EOP 1995
22	Reciprocating, screw, and scroll compressors; phaseout starts in 1996, EOP 2030
123	R-11 compressors, Low-pressure centrifugals; phased out in 1995
134a	First HFC, autos, domestic, chillers. New or retrofit CFC-12 compressors
404A	Long-term R-502 and R-22 alternative; new installations
408A	Interim retrofit solution for medium and low temperature refrigeration systems using R-502
409A	Interim alternative, (contains HCFC-22), R-12 transition refrigerant
502	Used in reciprocating compressors in food service; EOP 1995
507	New or retrofit applications designed to replace R-502
Greencool (Greencool, GU)	
405A	Interim alternative (contains HCFC-22); transition refrigerant for R-12
411A	Interim R-22 alternative
411B	Interim R-502 alternative
ICI (Arcton, KLEA)	
11	Most centrifugal compressors chillers; EOP 1995
12	Reciprocating compressors, some centrifugal compressors; EOP 1995
22	Reciprocating, screw, and scroll compressors; phaseout starts in 1996, EOP 2030
134a	First HFC, autos, chillers; new or retrofit R-12 compressors
407A	Long-term replacement for R-502; retrofit and new applications
407B	Close to existing R-502 conditions; retrofit and new applications
407C	Close to existing R-22 conditions
412A	Interim alternative; ultra-low-temperature R-500 alternative
502	Reciprocating compressors in food service; EOP 1995
508A	Low-temperature refrigeration
Indianapolis (GHGxx)	
12	Reciprocating compressors, some centrifugal compressors; EOP 1995
406A	Mobile air-conditioning; R-12 alternative

Table 2-25   Refrigerant Alternatives—Sorted by Manufacturer (continued)

Refrigerant Alternatives—Sorted by Manufacturer (continued)	
ASHRAE #	Notes and Uses
LaRoche	
11	Most centrifugal compressors chillers; EOP 1995
12	Reciprocating compressors, some centrifugal compressors; EOP 1995
22	Reciprocating, screw, and scroll compressors; phaseout starts in 1996, EOP 2030
134a	First HFC, autos, chillers; new or retrofit R-12 compressors
407A	Long-term replacement for R-502; retrofit and new applications
407B	Close to existing R-502 conditions; retrofit and new applications
407C	Close to existing R-22 conditions
412A	Interim alternative; ultra-low-temperature R-500 alternative
502	Reciprocating compressors in food service; EOP 1995
508	Reciprocating compressors in food service; phase out begins in 1996
Monroe Air Tech (GHG12)	
406A	Mobile air-conditioning; R-12 alternative
OZ Technologies (OZxx)	
12	Reciprocating compressors, some centrifugal compressors; EOP 1995
Rhone Poulenc (Isceon xx, Starton)	
403A	Interim alternative (contains HCFC-22); commercial/ transport refrigeration; transition for R-502
403B	Interim alternative (contains HCFC-22); transition for R-502

Table 2-25 Refrigerant Alternatives—Sorted by Manufacturer (continued)

REFRIGERANT SUPPLY AND DEMAND ISSUES

Refrigerant supply and demand issues are important considerations as you begin planning your options for managing refrigerants.

Demand for CFCs

Several sources have predicted a large demand for CFCs after the ban on production on January 1, 1996. The EPA warns that "rising costs and shortages of domestic and imported refrigerants can be expected" especially for R-12. By the year 2000, demand for CFC refrigerants could be as high as several hundred million pounds, and the demand will extend beyond the year 2000. It is also estimated that half of the R-12 produced in 1994 and 1995 will be used in automobile air conditioners.

One way to reduce demand for CFCs is through sound maintenance practices, especially by reducing the amount of CFCs lost due to leaking. Reports state that poorly maintained chillers leak at an average rate estimated at 15 to 25 percent. ASHRAE and other sources report that, based on historic refrigerant loss, from 20 to 40 million pounds (or more) of CFCs will be required annually just to replace the refrigerants lost through leaks and servicing.

EPA regulations say that if a chiller has annual leak rates greater than 15%, the owner must either repair, convert, or replace that chiller. The goal should be to reduce leaks to less than five percent per year which not only makes good economical sense but will reduce the overall demand for CFCs.

Note that new chiller design has reduced leak rates to below one-half of one percent. For a 500-ton R-123 chiller operating at that rate over a 30-year life, a lifetime replacement charge will be only about 12 gallons of refrigerant.

Supply and Availability of CFCs

So, what amount of CFCs will be available after the ban and what will they cost? Will there be enough CFCs for 1996 and beyond?

It depends on whom you talk to. In the years immediately following the production phaseout, most think CFCs will be generally available, but at a very high price. Others aren't so sure. Some sources are guessing they won't be available at any price.

But the majority seems to think the answer is no, there won't be enough CFCs after the production ban. The estimated demand for CFCs in 1996 of more than 40 million pounds tends to support that point of view.

On the other hand, supplies of R-22, R-123, and 134a will be adequate. Many manufacturers have invested in these products. According to reports, DuPont alone has adequate capacity of R-123 to supply the global needs for a low-pressure refrigerant, and supplies of R-22, the world's biggest seller, will be plenty.

Availability will depend, in large part, on how much CFCs are reclaimed. (Most reclaimed refrigerant is R-22.) A lot less reclaimed CFC is showing up than was expected by companies who got into the reclaiming business. No one is sure how much CFC refrigerant is being reclaimed. One industry estimate is 7.4 million pounds of CFCs would be reclaimed in 1993 and 1994. There are reports that some refrigerant reclamation businesses are obtaining large inventories of used refrigerant, but they are waiting to put them through the reclamation process after 1995 or later when CFC prices have gone up even more.

One possible reason more CFC refrigerant is not coming in for reclamation is that people are stockpiling their R-11 and R-12 and using it as part of their refrigerant management plan. The EPA is encouraging refrigerant stockpiling, but they don't have very good information on how many refrigerants are being stockpiled by owners as they retire old equipment.

Carrier, one of the major chiller manufacturers, has an offer that guarantees a supply of R-11 until the year 2000, for owners who add containment devices to their chillers. This offer, outlined in their "Chiller Operation Assurance Program," promises building owners that any manufacturer's brand of R-11 chillers "will operate until the year 2000, or Carrier will provide a replacement chiller free of charge." This offer of refrigerant supply can give some customers additional time to consider future retrofit or replacement plans with only a minimal initial expenditure.

Past Production Levels

Past refrigerant production levels can perhaps give an indication of future availability or shortages. CFC production overall has been decreasing, but the amount used for refrigeration has stayed about the same.

In 1993 R-12 sales were 189,000 metric tons, an increase over previous years. R-12 production peaked in 1989 at around 245,000 metric tons. Assuming levels of stockpiled R-12 are not significant, after production ceases, there will still be a high demand. Existing supplies will be used fairly quickly and prices will increase.

The production levels for R-12, R-11, and R-22 in the charts below are from the U.S. International Trade Commission, as reported in the April 11, 1994 *Air Conditioning, Heating & Refrigeration News.*

U.S. R-12 Production
(Millions of Pounds)

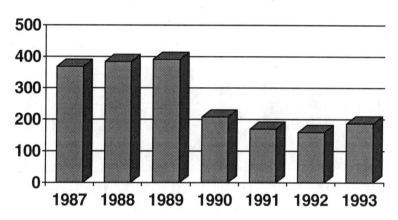

U.S. R-11 Production
(Millions of Pounds)

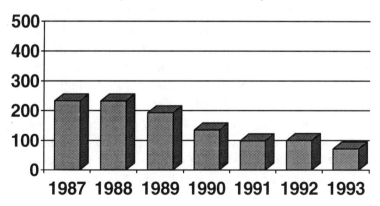

Figure 2-15 U.S. R-11 Production

R-22 production in 1994 was recently reported as 307 million pounds, a 5% increase over 1993. A production cap on R-22 will go into effect this year (1996) of approximately 220 million pounds (plus some additional allowance). R-22 use is increasing both in new construction and as a replacement for CFCs directly and as a component of many of the interim blends. The current low price of R-22 is a function of supply and demand. Most likely, the price of R-22 will rise.

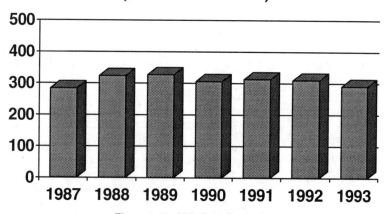

U.S. R-22 Production
(Millions of Pounds)

Figure 2-16 U.S. R-22 Production

R-11 and R-12 Price Projections

The following graph shows the projected future costs for R-11 and R-12 through the year 2020, and are from a study performed by Southern California Edison. The costs shown are representative of large chiller expenses. Costs for servicing small equipment such as residential refrigerators, is likely to be up to two times that shown in the graph.

The increases in cost for both R-11 and R-12 between now and shortly after the year 2000 reflects a period of decreasing supplies and continued demand. During this period CFCs will be available through reclamation only. As replacement of CFC equipment continues, demand for CFCs will continue to decrease until there is no longer any demand. At this point the CFCs will no longer be of value.

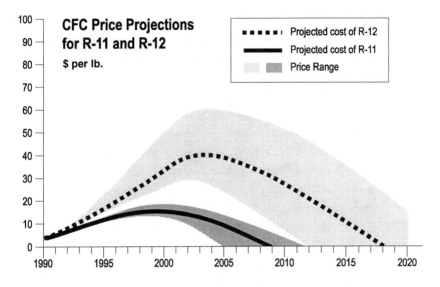

Figure 2-17   CFC Price Projections for R-11 and R-12

Availability of CFC Replacement Refrigerants

At the 1995 International Air-Conditioning, Heating, Refrigerating Exposition, in addition to the introduction of new refrigerants, the commercial availability of existing products was discussed.

AlliedSignal announced the world-wide availability of its Genetron AZ-20 (R-410A) and Genetron 407C (R-407C) refrigerants, both R-22 replacements.

The DuPont Company reports that R-407C is now commercially available to US original equipment manufacturers (OEMs) for new equipment, and will be available world-wide by the end of 1995.

ICI Klea announced that in April of 1995 they had begun the final stage in the expansion of production capacity for KLEA 134a (R-134a) at a plant in Louisiana. The plant capacity is currently at 40 million pounds per year and it will be expanded to more than 66 million pounds per year by early 1996. By 1996, ICI Klea's world-wide capacity for R-134a will be more than 120 million pounds per year.[38]

CHAPTER SUMMARY

Before the phaseout of CFC refrigerants began, there were only a few commonly used refrigerants and the choice you made was fairly simple. However, the changes in the rules that regulate refrigerant use and the dozens of newly developed refrigerants make it necessary to be better informed about refrigerants: their chemistry, how they are named, and the variables regarding how they are used. An improved awareness of these topics will help to minimize potential problems with equipment compatibility and will help maximize equipment safety.

Pure refrigerants, including ammonia, propane, and others, have been in use for a long time. Because of their minimal environmental impact, these have gained new interest and the exploration of new potential uses.

Utilization of these refrigerants and the various new blends present new challenges to owners and managers of businesses, system designers, and maintenance personnel. Ammonia's toxicity and propane's flammability are a concern. Azeotropes, and zeotropes are not only more expensive to make, but require special attention. These refrigerant blends have particular properties that present new challenges: new and converted systems must compensate for glide, fractionation, and in some cases higher operating temperatures and pressures, and will require focus on heat exchanger design, saturation pressures, tube design, and different lubricants. Maintenance issues include different system charging and refrigerant testing techniques, leak prevention and detection, and other issues.

More and more new refrigerants are appearing on the market as the rush to find alternatives continues. As more refrigerant testing and approvals occur and system innovations are developed, your options increase. Major manufacturers and small start-up companies alike are producing and promoting various new products. New systems and conversions represent a substantial investment, so you need to be cautious and make a well-informed choice.

If you are planning to maintain and preserve CFC-based systems as part of your management plan, a consideration of the supply and demand of CFCs, including R-11 and R-12, becomes important. Predicted shortages of R-12 may require conversion to another refrigerant, especially when the price of R-12 is driven beyond the conversion cost. R-11 also may experience some shortages until recovered R-11 becomes available after new chillers are installed. There may be enough R-11 to service existing machines for many years to come, especially if measures are taken to stop leaks. The cost of R-22 remains low for now, but its phaseout has begun. However, R-22 alternatives are becoming better understood and new equipment designed for R-22 alternatives is appearing in the market place.

Chapter 3

Air-Conditioning Machinery Choices with New Refrigerants

In this chapter we will discuss the various air-conditioning machinery choices available for consideration as you put together your refrigerant management plan. Chapter 4 takes a look at commercial and industrial refrigeration machinery choices with new refrigerants.

INTRODUCTION

In Chapter 1 we discussed the laws that regulate refrigerants and the reasoning behind them—ozone depletion, global warming, and safety concerns. In Chapter 2 we discussed CFC alternative refrigerants and answered common questions regarding replacement refrigerant properties, supply, future costs, and other topics.

So now, in light of these issues, how do you decide what equipment or refrigerant is best suited to your facility?

The first part of this chapter takes a look at the three options you have regarding what to do with your existing equipment—preserve and maintain existing equipment and CFC refrigerants, convert existing equipment to use different refrigerants, or replace existing equipment with new equipment. What you do with existing equipment relates to maintenance and making it conform to regulations. Making a decision to convert existing equipment or to purchase new equipment leads to considering refrigerant and machinery choices.

This chapter also will focus on the impact of different refrigerants on machinery design. Although a particular refrigerant's environmental impact may carry greater weight in choosing air-conditioning machinery today than it used to in the past, the overall equipment choice involves many additional factors. Each individual application has a unique set of criteria: cost, availability, efficiency, and environmental impact. However, the refrigerant choice does dictate much of the machinery design, and each refrigerant has its own advantages and disadvantages.

The impact of a particular refrigerant on machinery design specifically relates to types of heat exchangers and their design, types of compressors and their design, and for centrifugal chillers, the advantages and disadvantages of different operating pressures.

OPTIONS FOR EXISTING AIR-CONDITIONING EQUIPMENT

As you make decisions on how to react to new and future regulations for air-conditioning equipment, there are three basic paths you can take:

• Preserve existing equipment and CFC refrigerant.
• Convert existing equipment and refrigerant.
• Scrap what you have and buy new equipment.

Air-conditioning equipment can last as long as 30 years, so this is not a decision you make lightly since there will be long-term economic and environmental consequences. Most governing agencies and professional organizations recommend an action plan to help you make this decision.

The action plan entails analyzing a comprehensive set of selection criteria—one "size" doesn't fit all. All of the equipment and refrigerant choices available need to be considered plus a wide range of factors including operating costs, any utility rebates that might be available, total system energy efficiency, refrigerant availability, and any change in capacity you might need if there are plans to remodel the building. (We'll explore this in more detail in Chapter 5.)

Option 1: Preserve Existing Equipment and CFC Refrigerant

If you determine that your current equipment has some life left, and refrigerant supply is adequate, you need to act on three key issues regarding preserving your current CFC refrigerant:

- Improve refrigerant containment.
- Improve equipment maintenance.
- Follow sound servicing practices.

As long as there is a supply, using CFCs is still legal. What's not legal is mishandling CFCs. One source of trouble is leaking refrigerants. If you allow refrigerants to leak, it is not only illegal but expensive (Section 608 of the Clean Air Act is discussed in Chapter 1). This issue is solved by taking steps to meet containment requirements.

The first containment measure to take is repairing all known refrigerant leaks, including refrigerant leaking out and air "leaking" in.

Next, more efficient purge equipment needs to be incorporated into your existing low-pressure chiller systems. Since most centrifugal chillers that use CFCs are low-pressure machines, the vacuum produced inside can draw unwanted air into the system. A purge system is used to remove the air. In a typical purge process, the air is released and along with it, refrigerant. A system that is not "tight" will allow in more air, which means purging is required more often. (Purge systems are not an issue for medium- and high-pressure systems.)

This loss of refrigerant during the purge process can be substantially reduced by using high-efficiency purge equipment now available. Some purges on today's market do not allow (essentially) any refrigerant to escape. As one example, a purge manufactured by Trane allows less than .0049 pounds of refrigerant per pound of dry air to escape. For a typical 500-ton centrifugal chiller, this amounts to less than three quarters of an ounce per year. This purge also incorporates a microprocessor-controlled runtime meter that records and displays the amount of time the purge operates. The "minutes of operation" can be used to indicate whether a machine is "tight" or not.

By monitoring refrigerant levels with highly accurate sensors, it is possible to provide an extra margin of safety for all refrigerants. Using sensors to monitor leaks meets the requirements of ASHRAE Standards 15-1992 and 15-1994. The sensors measure low ppm concentration of refrigerant vapors and reduced oxygen levels. (A discussion of implementing a leak-detection system is found in Chapter 5.)

One way to reduce leaks in and out of an air-conditioning system is proper equipment maintenance. This involves the simple task of tightening threaded connections on the chiller on a regular basis.

Finally, to improve refrigerant containment, follow sound servicing practices. There are new regulations regarding the legal way to service a chiller. Not only will proper servicing techniques reduce refrigerant losses, it will, along with proper documentation and good operating practice, avoid legal headaches. In Chapter 5, these issues are discussed in greater detail.

Option 2: Convert Existing Equipment

If you have decided that your existing equipment and CFC refrigerant cannot be preserved, the second option for existing air-conditioning systems is to convert to a non-CFC refrigerant. This usually involves a change to an HCFC or HFC refrigerant. Typical conversions are to R-123 as a replacement for R-11, and to R-134a for R-12 and R-500. Converting a chiller to run with these substitutes usually requires some modification to the compressor or lubricant used.

Converting a system to a non-CFC refrigerant is comparatively more expensive than preserving and maintaining your existing equipment and refrigerant. In addition, a converted chiller will often lose cooling capacity and may use more energy. But don't be discouraged by these realities: it's very possible that your existing system is oversized, as many existing systems are, so the loss of capacity may not present a problem. Also, a conversion plan will also explore other areas that may identify energy-saving opportunities, such as lighting conversions, insulation, and control systems that can reduce your overall cooling load.

Conversion choices are often made based on:

- A corporate response to environmental regulations.
- A response to perceived or real shortage of CFC refrigerants.
- A total loss of CFC refrigerant.

The choice to covert to a non-CFC refrigerant can in some cases be driven by corporate policy (either political or philosophical) to make an environmental statement and eliminate all CFCs from your facility. This means a choice to convert may not be made solely on the short-term economics of preserving an existing chiller system.

On one hand, it may seem practical to keep using CFCs and preserve the existing equipment when the microeconomics of the chiller system are considered separately—its related operating expenses, remaining life, payback period, return on investment, etc.

On the other hand, often the total economics of the enterprise (macroeconomics) may drive the decision. For example, some companies that adopt conscientious environmental policies and earn the recognition of being an Energy Star organization may be in a better position to qualify for profitable government contracts, which might provide revenue many times the savings associated with retaining the existing chiller system or many times the costs of converting. So a decision to change may be made, not only on the chiller itself, but based on other philosophical or political reasons.

Others have made conversion decisions because they are concerned about running short of a phased-out refrigerant. Although good planning may help avoid an immediate shortage, refrigerants that are being phased out will be scarce at some time in the distant future.

Another occurrence that would present an opportunity to decide to convert is a total loss of CFC refrigerant due to a catastrophic failure. If for some reason a malfunction caused the loss of all refrigerant in a chiller, a change at that time to a non-CFC refrigerant may make sense economically, since non-CFC refrigerants are often less expensive than replacing CFCs.

For example, if an average R-11 system with a 1500 pound refrigerant charge should experience a complete loss of refrigerant, the cost to replace the charge would be about $15,000 at current prices. R-123, a typical conversion refrigerant, would cost about $6,000 for a complete system recharge. The difference in refrigerant cost could be applied toward the cost of a conversion to R-123.

Option 3: Buy New Equipment

The third option is to plan for purchasing a new chiller and the required additional system components. If the existing equipment is at or near the end of its useful life, chances are the chiller is not exceedingly efficient. So, instead of spending money on conversion or improved containment of the old machine, a new system may be the best option.

But you need to consider that buying new chiller machinery sometimes triggers remodeling the entire central plant including adding new cooling towers, a secondary pumping system, or perhaps a thermal energy storage system.

In the recent past, the choice of refrigerant was a minor factor in the selection of new air-conditioning machinery. Today, when choosing a new chiller, the refrigerant choice is a key deciding factor, in addition to a variety of other "traditional factors" such as:

- Configuration of the chiller plant, including the number of chillers required for the total cooling load.

- Initial cost.

- Energy efficiency and energy cost.

- Summer peak energy use.

- Manufacturer preference.

The new refrigerants may have a significant impact on machinery choice, including chiller design, efficiency, and type of compressor. The rest of this chapter discusses the impact of new refrigerants on chiller design.

NEW REFRIGERANT IMPACT ON THE DESIGN OF AC EQUIPMENT

To provide context for later, we'll start our discussion on the impact of new refrigerants on the design of air-conditioning equipment with a very brief overview of the basic components of a water-cooled chiller system and explain a little about how a chiller works.

A Typical Water-Cooled Chiller System

The diagram on the next page shows a typical chiller system that operates on a vapor compression cycle. This basic system has a chiller to create chilled water, chilled water distribution system and air handling units to distribute cooled air, and a cooling tower to dispose of system heat.

The basic components of a chiller are the compressor, condenser, evaporator, flow control valve, and the refrigerant. There are other types of refrigeration systems such as absorption or steam-jet, but our discussion will focus on vapor compression since this type is impacted by the phaseout of CFCs.

In vapor compression refrigeration, the refrigerant is used to alternately remove heat from one location and reject that heat to another location. The process repeats and the refrigerant can complete the cycle from cold to hot over and over, many million times in the life of the system, and the refrigerant doesn't wear out. The compressor may fail or the refrigerant may leak out, but it doesn't wear out.

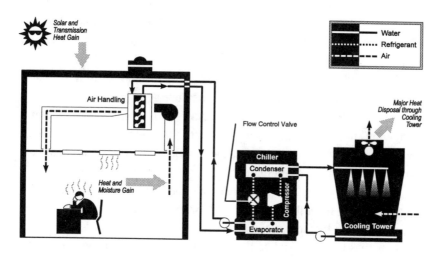

Figure 3-1 Example of a Water-Cooled Chiller System

As shown in the diagram above, heat inside a building causes the air to warm up. The warm air is blown by a fan over a cooling coil that has cold water circulating through it. The water in the coil warms up and the air cools down.

The water makes its way to the chiller evaporator. It is known as an evaporator because liquid refrigerant evaporates there ("boils" and undergoes a "phase change" to a gas) and removes heat from the returning water and produces chilled water.

Chiller evaporators are either metal tubes or plates that have refrigerant on one side and water on the other. Heat is easily transferred across the metal tubes of the evaporator without the water and refrigerant actually coming into contact.

The evaporated refrigerant or gas is drawn from the evaporator into the compressor which reduces its volume many times; as a result of this compression, the refrigerant pressure increases and the gas heats up. It takes work to turn the compressor, which is usually supplied by an electric motor.

After passing through the compressor, the hot gas enters the condenser, another metal heat exchanger of tubes or plates, where the refrigerant condenses from vapor to liquid. As it condenses, the heat is transferred to the cooling tower water. The cooling tower water disposes of the heat to the outdoor air through the evaporation of water (like an evaporative cooler).

The condensed or liquid refrigerant is still at a high pressure because the compressor maintains the pressure as long as it is operating. The flow-control valve regulates the flow of the high pressure liquid refrigerant from the condenser back into the evaporator which is at lower pressure. As long as it runs, the compressor creates and maintains a low pressure in the evaporator.

The liquid refrigerant evaporates at this low pressure because the water returning from the cooling coil is warmer than the refrigerant. The refrigerant absorbs the heat, changes from liquid to gas and then enters the compressor again. And so the refrigerant continues its cycle until the compressor is turned off and all the pressures equalize throughout the system. Some refrigerants have been operating in systems for more than 50 years, and as long as the other systems' components don't fail, the refrigerant won't either.

Heat Exchanger Flow Design

In a conventional, large-scale air-conditioning system, there are several heat exchangers. The evaporator and condenser are of most interest in terms of new design required by new refrigerants. (Other heat exchangers include the air-cooling coil at an air-handling unit and the cooling tower.) Given a specific type of compressor and refrigerant, chiller efficiency is to a great extent a function of the size of heat exchangers. Larger heat exchangers will improve efficiency but they also increase the cost.

Cross Flow Heat Exchanger

With pure refrigerant compounds such as R-12, the flow pattern of the refrigerant and water in the heat exchanger (condenser or evaporator) was not a significant concern because the temperature of the refrigerant remained the same throughout the entire circuit. As a result, heat exchangers were designed with cross flow circuits since this is the least expensive design to manufacture.

In a cross flow heat exchanger, the refrigerant flows in a horizontal circuit and the water flows vertically or diagonally across the tubing that contains the refrigerant flow. Since the refrigerant temperature is the same at all places in the refrigerant tubing, the flow pattern of the water does not consider whether the refrigerant just entered the heat exchanger or is exiting.

The cross flow design requires larger heat exchangers and typically provides a less efficient heat exchange.

In Chapter 2 we mentioned that for some newer refrigerant blends while under constant pressure, the evaporating and condensing temperatures change as the blend composition shifts (the amount of the individual blend components that are liquid and gas changes). This temperature change during a constant pressure phase change is called *glide*.

With zeotrope temperature glide, the temperature of the refrigerant at one end of the phase change process (at the evaporator or condenser) is many degrees different than at the other end. (With R-407C, for example, the temperature difference is about 12° F.)

Because blends behave differently, and glide makes cross flow design less efficient, they will require different flow patterns for the water inside the heat exchanger. Since most condensers and evaporators today have a cross flow design, a converted system will more than likely experience a loss of capacity.

The following graphic shows the cross flow paths of water and refrigerant through an evaporator for a typical refrigerant that exhibits glide. The indicated points on the graphic show typical water and refrigerant temperatures during the heat exchange. The key aspect demonstrated by the graphic is the difference in temperatures between the refrigerant and water (Δt). This varying temperature differential makes the heat exchange less efficient.

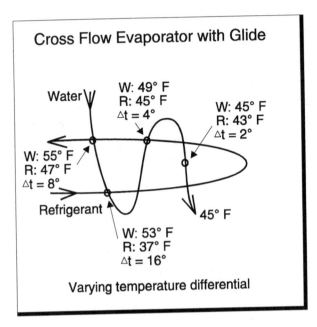

Figure 3-2 Cross Flow Evaporator with Glide

Because of the temperature differential, a better heat exchanger design for zeotropes with high glide is counterflow.

Counterflow Heat Exchanger

In counterflow heat exchangers, the water and refrigerant flow in opposite directions.

The graphic in Figure 3-3 shows the changing temperature of water and a refrigerant with glide in an evaporator and indicates water flowing one direction and refrigerant the opposing direction. Heat is transferred from the water to the refrigerant. The water enters the evaporator at about 55 degrees (A) and is cooled to about 45 degrees (B). The refrigerant enters the evaporator at about 37 degrees (C) and exits warmed to about 47 degrees (D). In the course of these temperature changes, the refrigerant goes through a phase change from liquid to gas. A pure refrigerant does not have glide and remains at a constant temperature (C to E) in the evaporator.

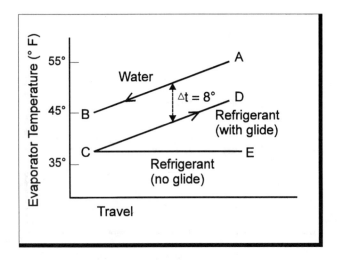

Figure 3-3   Counterflow Evaporator Temperatures

In the figure above, the difference in temperature (Δt) is measured as the vertical distance between the water diagonal line and the glide refrigerant diagonal line. In this case where the two fluids change temperature, a counterflow design helps keep the difference in temperatures between the water and refrigerant almost constant throughout the entire trip through the evaporator. This temperature differential is important because it assures an even heat exchange throughout. Also, the temperatures of the water and refrigerant as they leave the heat exchanger can be brought closer to their "ideal" values, which improves efficiency.

Figure 3-4 on the following page again demonstrates a counterflow design with the refrigerant and the water flowing in opposite directions inside the heat exchanger. The entering refrigerant (at its coldest for an evaporator) has a heat exchange with the coldest water (the water leaving the evaporator). Likewise, the warmest refrigerant (leaving the evaporator) absorbs heat from the warmest water (the entering water). This graphic also shows how counterflow keeps a constant temperature differential throughout the evaporator.

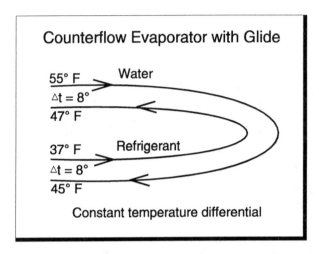

Figure 3-4   Counterflow Evaporator with Glide

For condensers, the counterflow heat exchangers release heat; the warmest water exchanges with the warmest refrigerant.

True counterflow heat exchangers are not yet commercially available. However, this design is considered both possible and practical, and some manufacturers are presently experimenting with designs. In theory, counterflow design can actually optimize the effectiveness of heat exchangers when temperature glide is a factor, but they will probably be larger and more expensive to manufacture.

Note that the previous discussion about counterflow heat exchangers is limited to air-conditioning systems that use an expansion valve for metering refrigerant to the evaporator, typically reciprocating and screw chillers. In contrast, centrifugal chillers typically use a "flooded" shell-and-tube evaporator. In this type, the water flows through tubes inside the evaporator shell that is filled (or "flooded") with refrigerant. The refrigerant is maintained at a specific level in the evaporator with a float valve. Because of this arrangement, it will probably be difficult to use a zeotrope refrigerant with a flooded evaporator since there are varying boiling temperatures and a non-constant heat exchange. For these reasons, centrifugal chiller system conversions that retain flooded evaporators may require pure refrigerant alternatives.

Compressor Design

The compressor is essentially a pump that maintains the refrigerant in the evaporator at a pressure and temperature that is low enough to remove unwanted heat, and then increases refrigerant pressure and temperature enough to reject heat at the condenser. Some heat is added by compressor friction and this is also rejected at the condenser.

Hermetic compressor designs use an enclosure around the compressor and the electric motor drive. In this type of design, the motor is cooled by refrigerant that circulates around the motor windings. With some of the new refrigerants and their associated lubricants, steps must be taken to make sure the motor is compatible to avoid damage. Because they are cooled by refrigerant, hermetic compressor motors are usually smaller for a given horsepower; however, the additional heat absorbed by the refrigerant must also be rejected to the condenser.

External or open drive compressor designs use an electric motor positioned outside of the compressor casing. This makes the motor more accessible. The motor heat is not absorbed by the refrigerant but is rejected to the surrounding ambient air. This type of design requires a seal around the drive shaft to prevent the refrigerant and oil from leaking out.

As we'll see, there are several types of compressors that are used in chiller systems. All of them accomplish the same objective: to increase the pressure and temperature of the refrigerant as it is transferred from the evaporator to the condenser. This increase is also called "lift," derived from lifting the low evaporator temperature to a high condenser temperature.

Although the refrigerant is not necessarily the primary focus when choosing a particular system, the selected refrigerant does dictate the design, efficiency, and type of compressor used in a chiller.

We'll continue our examination of air-conditioning equipment with a brief discussion of types of chillers based on compressor design including:

- Reciprocating compressors.
- Screw compressors.
- Scroll compressors.
- Centrifugal compressors.

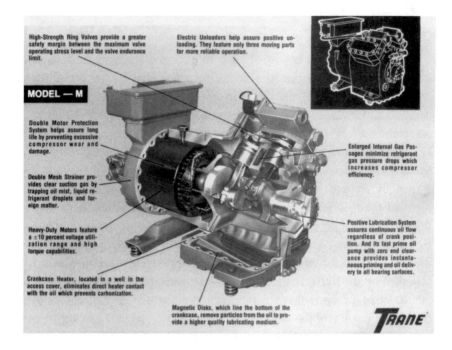

Figure 3-5 Reciprocating Compressor

Reciprocating Compressors

Reciprocating compressors use positive displacement by means of pistons to compress the refrigerant. The pistons move inside cylinders to reduce the volume of the vapor compression chamber.

A reciprocating compressor may have one or more pistons. For example, a 150-ton chiller may have four compressors with four pistons each. The pistons are usually driven directly through a pin and connecting rod from the crankshaft that is turned by an electric motor.

Since each cylinder handles a fixed volume of refrigerant gas, building larger reciprocating chillers is a matter of adding cylinders or compressors. Reciprocating chillers tend to be most cost effective at smaller loads.

Reciprocating compressors are used with small residential air conditioners and rooftop package units, as well as for small commercial systems. This type of chiller typically is used in systems where first cost, available floor space, and installation cost are more important issues than operating costs. For sizes between 10 and 200 tons of refrigeration capacity, they make up a significant portion of the chiller market.

The reciprocating compressor maintains a constant volume flow rate over a wide range of pressure ratios. An efficient reciprocating compressor, at full load, can achieve an efficiency of 0.8 kW/ton. Smaller ones, up to about 15 tons, average 0.9 kW/ton efficiency. (kW/ton is a method of expressing chiller efficiency. It is equal to the kW needed to operate the compressor's motor divided by the tons of cooling the chiller delivers. The lower the kW/ton, the more efficient the chiller.)

One drawback of reciprocating compressors is that they have many moving parts, and there is friction between all of them. This means these machines require higher maintenance and are less reliable because of the potential for component failure. Also, to maintain efficiency, tight clearances are required between the pistons and cylinder walls.

Refrigerant Alternatives For Reciprocating Compressors in Air Conditioning

The refrigerant used almost exclusively in reciprocating air-conditioning compressors has been the pure refrigerant R-22. However, R-22 is considered a "transitional alternative" or interim refrigerant because it contains chlorine.

This segment of the market, which accounts for almost one-third of the installed chiller capacity and two-thirds of the annual shipments of packaged chillers is concerned with discovering appropriate alternative refrigerants.

Because R-22 has high-density molecules, this refrigerant has good heat transfer properties which reduces the size of heat exchangers and compressors.

For converted reciprocating chiller systems, the best alternatives are single-component refrigerants, azeotropes, or zeotropes with very low glide.

Alternative refrigerants for reciprocating chillers include R-410A (an azeotropic mixture of R-32 and R-125), R-407C (a zeotropic blend of R-32/R-125/R-134a), and R-134a (a pure fluid). Other pure fluid alternatives for R-22 such as ammonia, and propane also are being considered as alternatives.

R-410A exhibits azeotropic properties and has virtually no temperature glide. Currently, there are no machines manufactured that use R-410A or R-410B. With this refrigerant, heat exchangers won't need a counterflow design.

However, R-410A has a very high discharge pressure (more than 400 pounds per square inch, or approximately 50% higher than R-22 at a given temperature) and will need new system designs to tolerate the elevated pressure. This high pressure leads to better performance in heat transfer and actually lowers overall temperature lift at the compressor, but the pressure increase has an impact on the high-pressure components of the system, namely the compressor, condenser piping, tubes, valves, etc. Optimal design may render smaller but stronger system components, but existing systems would not be able to tolerate the additional stress from the higher pressure.

R-407C operates at about the same pressure and transports a similar amount of heat for a given mass flow as R-22. However, this blend exhibits a temperature glide of 12° F.

True counterflow heat exchangers can help R-407C perform better, perhaps better than R-22. Existing R-22 reciprocating compressors should be able to use R-407C, but most likely won't be able to use R-410A.

Air conditioning equipment manufacturers will not necessarily want to change to R-134a. Because R-134a has poorer heat transfer properties, it requires a greater heat exchanger surface area, physical size, and as a result, higher cost.

Screw Compressors

Screw (or helical rotary) compressors, like reciprocating units, are positive displacement machines. They have either two matched spiral-grooved rotors, (intermeshing "screws") or one rotor with a "gaterotor." As they turn, the volume of the refrigerant chamber between the screws is reduced, compressing the refrigerant.

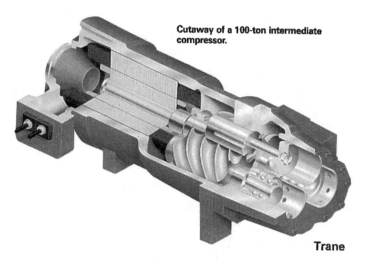

Cutaway of a 100-ton intermediate
compressor.

Trane

Figure 3-6   Screw Compressor (1)

End view showing male and female
rotors and slide valve on an 85-ton
intermediate compressor.

Trane

Figure 3-7   Screw Compressor (2)

The screw compressor has nearly constant-flow performance and can provide high-pressure ratios. The flow of gas in the rotors is both radial and axial.

Screw compressor liquid chillers are available as package units from about 15 tons to 850 tons in both open and hermetic styles. Package units with "receivers" (not water-cooled condensers), are made for use with air-cooled or evaporative-cooled condensers. Most factory-assembled liquid chilling package units use R-22 or R-134a. R-407C or R-410A may eventually replace R-22, commonly used with screw compressors.

Screw compressors are up to 40% smaller and lighter than centrifugal compressors of the same cooling capacity. Screw compressors typically operate at 0.63 to 0.80 kW/ton. System efficiency requires tight clearances between screw lobes.

While screw compressors are available from 15 to 1,250 tons, their greatest use is in systems from 100 to 700 tons. In this size range, they make up about 10% of the chiller market. Demand is growing for today's most efficient screw chillers.

Scroll Compressors

Scroll compressors are rotary motion positive-displacement machines. They are found in small units used for domestic air conditioners and in the small commercial market. They range in size from 1.5 to 15 tons.

This type is similar to screw compressors, but the refrigerant is compressed by using two meshed, spiral-shaped scroll rotors. Typically, one rotor is fixed and one is moveable.

The refrigerant vapor is compressed by continually reducing the size of the refrigerant chamber. The suction gas is sealed in "pockets" at the outer edge. The gas is compressed as the size of the pockets is progressively reduced as the scroll motion moves the pockets inwards towards the discharge port. Several pockets of refrigerant are compressed at once which keeps the pressure profile of the output relatively even. The moveable scroll rotates in "orbits" which is a sequence of suction (inlet), compression and discharge phases.

Scroll compressors are, in general, more efficient than other small compressors. They have few moving parts (causing less friction) and are smaller than reciprocating compressors. However, system efficiency

One of two matched scroll plates —
the distinguishing feature of the scroll
compressor.

Trane

Figure 3-8 Scroll Compressor

requires tight clearances between scrolls. There are no valves, so there is no refrigerant lost through the valves. The compression chamber empties completely, so there is no re-expansion volume as with reciprocating compressors. The intake and discharge openings are physically separated, so the heat of compression stays with the discharge gas rather than feeding back into the intake gas. These features result in efficiency ratings up to 83%, which is comparable to larger screw or centrifugal units.

Scroll compressors will typically use R-22, and R-407C is a future alternative. Scroll compressors may be good candidates for a more "hefty" design to incorporate high-pressure refrigerants such as R-410A and R-410B. If provided with an adequate evaporator and condenser, blends with high glide may be used with scroll compressors.

Centrifugal Compressors

Centrifugal compressors use centrifugal action to compress the refrigerant. A centrifugal compressor, similar to a pump or a fan, compresses the vapor flowing through it by spinning it from the center of an impeller wheel, radially outward at high speeds.

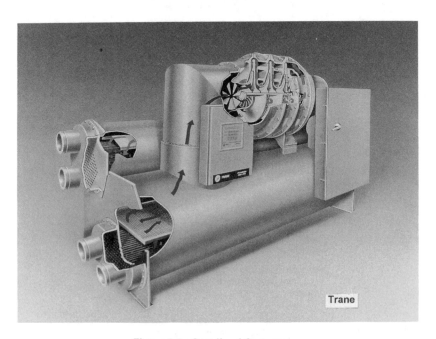

Figure 3-9 Centrifugal Compressor

Centrifugal compressors have very few moving parts. This reduces maintenance and increases reliability.

Since the centrifugal compressor is not a constant-displacement type, it offers a wide range of capacities and handles varying refrigerant flows easily. By changing built-in design items (including the number of stages, compressor speed, impeller diameters, and the refrigerant used), it can be used in chillers that have a wide range of chilled-liquid temperatures. Because this type can vary capacity continuously to match a wide range of load conditions (with nearly proportionate changes in power consumption), it's a good choice for both close temperature control and energy conservation.

Centrifugal chillers are an excellent choice for larger cooling loads. In 1990, there were more than 100,000 installed throughout the world, primarily used for cooling commercial, industrial, and large public buildings. From 200 to 10,000 tons, they make up more than 90% of the chillers purchased. However, for smaller loads, high compressor costs makes a centrifugal chiller less cost effective. Centrifugal efficiency levels today are in the range from .50 to .75 kW/ton and are the most energy-efficient means of air conditioning available.

Low-pressure centrifugal chillers have been popular in the past due to their efficiencies and reliability. Medium- and high-pressure centrifugal chillers have traditionally been used for larger capacities to keep size and cost down.

The following table shows the approximate number of centrifugal chillers in service, listed based on the different refrigerants they use.

Approximate U.S. Centrifugal Chillers in Service (early 1990s)				
Refrigerant Type	Number of Units	% of Total Units	Average charge (lbs.)	Refrig. in Service (million lbs.)
R-11 (CFC)	64,000	80	1100	65.1
R-12 (CFC)	8,000	10	1600	11.8
R-113, 114, 115 (CFCs)	800	1	800	0.9
R-22 (HCFC)	3,000	4	4000	11.8
R-500 (CFC)	3,700	5	2600	9.6
Total	79,500			99.2

Table 3-1 Approximate U.S. Centrifugal Chillers in Service (early 1990s)

Centrifugal Chiller Efficiency and Costs

One effective way to compare centrifugal chillers on the market is by using kW/ton efficiency ratings. Chiller cost and efficiency will vary depending on the refrigerant used.

Given a specific type of compressor and refrigerant, chiller efficiency is primarily a function of the size of heat exchangers. Larger heat exchangers will improve efficiency but they also increase the cost. But no matter how large you make the heat exchangers, there is a limit to how much they can improve efficiency since the evaporating temperature is limited to that of the leaving water temperature. For a centrifugal compressor, chiller system efficiency also depends on operating pressures, impeller design and rotation speed, and the number of stages of compression. kW/ton is a function of and is limited by the "ideal cycle" efficiency of the refrigerant and compressor efficiency.

Ideal Cycle Comparison—Centrifugal Chillers (40°F Saturation Suction 100°F Saturation Discharge; No subcooling or superheat)						
Refrigerant	Evaporator Pressure (psia)	Condenser Pressure (psia)	CFM/Ton	COP	kW/Ton	% of R-11
R-11 (CFC)	7.02	23.46	15.85	7.570	.465	100.0
R-123 (HCFC)	5.78	20.77	18.85	7.435	.472	98.2
R-114 (CFC)	15.08	45.85	8.99	6.864	.512	90.7
R-12 (CFC)	51.67	131.86	3.06	7.061	.498	93.3
R-500 (CFC)	60.72	155.80	2.62	6.702	.525	88.5
R-134a (HFC)	49.76	138.90	2.99	6.937	.507	91.6
R-22 (HCFC)	83.21	210.60	1.91	6.984	.503	92.3
R-717 (ammonia)	73.11	211.40	1.70	7.261	.484	95.9
R-290 (propane)	78.78	189.04	2.26	6.851	.513	90.5

Trane

Table 3-2 Ideal Cycle Comparison (Centrifugal Chillers)

When comparisons are based on ideal properties, low-pressure centrifugal chillers can provide the best design efficiencies since R-11 and R-123 have better cycle efficiency. Chillers that use the higher pressure refrigerants R-22 and R-134a generally have better refrigerant-side heat transfer properties. R-22 reciprocating chillers have relatively poor efficiency and are sold in low-cost package units that do not allow for optimum heat exchanger sizing.

The latest price quotes from centrifugal chiller manufacturers range from $150 to $300 per ton. High-efficiency machines can be up to 25% more expensive than standard-efficiency machines.

Operating Pressures

One way chillers can be classified is by the pressure at which they operate: low, medium, or high pressure. Which type is best? Again, the answer depends on many factors.

The table below shows typical chiller operating pressures for some of the refrigerants we have been discussing.

Operating Pressures				
Operating Pressures	Refrigerant	Evaporator (at 38° F)	Condenser (at 100° F)	Unit Pressure when Off-line (at 72° F)
Low	R-123	-18.7 in Hg	6.1 psig	-5.6 in Hg
	R-11	-16.3 in Hg	8.8 psig	-1.6 in Hg
Medium	R-134a	33.1 psig	124.1 psig	74.0 psig
	R-12	35.2 psig	117.2 psig	72.9 psig
High	R-22	65.6 psig	195.9 psig	125.7 psig
	R-407C	70.3 psig	204.9 psig	139.3 psig
	R-410A	115.3 psig	316.9 psig	204.3 psig
in Hg = inches mercury (vacuum) psig = pounds per square inch gauge				

Table 3-3 Operating Pressures

Low-Pressure Centrifugal Chillers

Low-pressure (or "negative-pressure") chillers typically operate at less than 30 psi (pounds per square inch). (The condenser typically operates above atmospheric pressure and the evaporator below.) This type uses R-11 and R-123 refrigerants, which are used only with low-pressure compressors.

There is potential for the refrigerant to leak out of low-pressure systems, or for air to leak into the system. When air is drawn into the system, low-pressure chillers require purge systems to remove the air and other non-condensable substances. We discussed earlier that as the air is purged out of the system, typically some refrigerant also is removed and lost.

The low-pressure refrigerants R-11 and R-123, because of their larger molecule size, have lower density. This means chillers that use these refrigerants typically are larger than higher pressure machines. The low difference in pressure between the evaporator and condenser minimizes internal leaking between these components, which results in good compressor efficiency. The maximum working pressures are low enough to be exempt from the ASME (American Society of Mechanical Engineers) construction code, which results in lower system cost.

Most low-pressure machines use hermetic motors, where refrigerant is used to cool the motor, but some use open motor drives that are cooled by ambient air.

As with any system, low-pressure chillers have advantages and disadvantages.

Advantages and Disadvantages of Low-Pressure Chillers	
Advantages	**Disadvantages**
• Reliability (no gears required) • Efficiency (no gear and additional bearing loss; low velocity gas, less friction; Multi-stage design can utilize economizer) • ASME pressure rating not required (lower costs)	• Larger compressor size (greater cost) • Larger heat exchanger size (greater cost) • May require multi-stage impellers (greater cost) • Requires air purge (some refrigerant loss)

Table 3-4 Advantages and Disadvantages of Low-Pressure Chillers

Low-Pressure Refrigerants

Because R-11 is a CFC and is among those refrigerants being phased out, a suitable alternative for R-11 is sought. For the near-term, R-123 is the leading alternative for R-11.

The reason R-11 was the most commonly used refrigerant is that it had, and still has, the highest cycle efficiency of any refrigerant in use today for low-pressure chillers. Machines that use R-123 have efficiency ratings as low as .50 kW/ton. R-123 toxicity levels are acceptable by the EPA and R-123 is widely available and is being manufactured in large quantities.

However, R-123 does contain chlorine and a total production phase-out of R-123 is scheduled by 2030. But during the next 35 years or so R-123 can be used, which is about the average lifetime of new equipment. A better alternative would be one without chlorine or bromine and zero ODP.

Centrifugal chillers specifically designed for naval vessels use R-114, which is another low-pressure refrigerant. The capacity range for these chillers is from 125 to 400 tons. The boiling point of R-114 is very close to atmospheric pressure at typical evaporator temperatures. This attribute combined with a design with positive pressure at the evaporator prevents moisture-laden sea air from leaking into the system and helps avoid potential corrosion problems. R-114 is a CFC and was phased out in 1995.

One future alternative for R-11 and R-123 is the chlorine-free low-pressure refrigerant R-245ca. In preliminary tests, R-245ca has been shown to be a good far-term replacement candidate. The thermophysical properties of this compound are very similar to those of R-11 and R-123. Also, R-245ca has an ODP of zero and a GWP about one-third that of R-134a. Toxicity testing of R-245ca has been very limited and conclusive results are not available; so far, toxicity levels are low. Flammability may be a minor issue but if R-245ca is used in a blend this issue could be resolved.

However, because R-123 is competitively priced and has a very low ODP and GWP, and has near zero refrigerant loss, the development of other R-11 replacement refrigerants is a lower priority for manufacturers.

Medium- and High-Pressure Centrifugal Chillers

Medium- and high-pressure ("positive-pressure" or "positive-displacement") centrifugal chillers operate at greater than 30 psi (pounds per square inch) which is more than atmospheric pressure. Both evaporator and condenser operate above atmospheric pressure.

Medium-pressure chillers use R-12 and R-134a refrigerants, and the high-pressure machines use R-22 especially in the largest centrifugal chillers (above 2000 tons).

Medium- and high-pressure centrifugal chillers are usually smaller in size because they use denser refrigerant gas which requires smaller compressors and heat exchangers. Medium-pressure machines that use R-12 and R-134a refrigerants operate at higher speeds and need a gear drive. Most use hermetic motors, but some open drives are also used. The moderate refrigerant density and moderate pressure differential results in practical, high-speed, single- and two-stage impeller designs.

Because these chillers operate under greater pressure with higher densities, they require higher compressor speeds and thicker construction materials, which increases their cost. The chiller shells must be certified by the American Society of Mechanical Engineers (ASME), and are inspected during production. They are designed with features such as joints using hydraulic O-ring fittings which help reduce leaks. However, large high-pressure centrifugal chillers that use R-22 are typically smaller and less expensive when compared to other systems. No purge is required and no external storage tanks are required for servicing since the refrigerant charge can be isolated internally.

Medium- and High-Pressure Refrigerants

R-12 and R-134a are used in medium-pressure centrifugal chillers, and R-22 is used in high-pressure chillers. (R-12 and R-22 alternative refrigerants are discussed in detail in Chapter 2.)

R-134a chillers are typically less efficient, but improved heat exchanger surfaces compensate for this drawback. Some R-134a chillers incorporate a turbine to recover energy that would otherwise be lost in the expansion portion of the vapor compression cycle. The turbine, attached to an extended motor shaft, uses the pressure difference that exists between the condenser and the evaporator to supplement the ener-

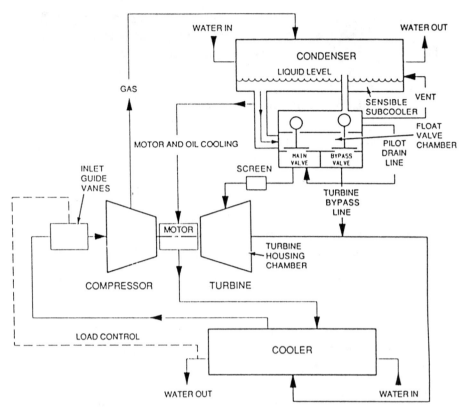

Figure 3-10   Carrier 19XT Refrigeration Cycle

gy furnished by the chiller motor and reduces external energy requirements. (See the figure above.) The result is energy efficiencies in the range of .56 to .59 kW/ton.

R-500 (a mixture of 73.8% R-12 and 26.5% R-152a) is sometimes used in converted R-12 centrifugal chillers to increase the capacity range.

The medium- and high-pressure chiller systems have their own advantages and disadvantages.

Advantages and Disadvantages of Medium- and High-Pressure Chillers	
Advantages	**Disadvantages**
• Smaller size for all components (less cost) • No purge equipment (less cost)	• Need gear drives (additional energy losses and components; higher costs) • ASME pressure rating required (higher costs) • Only one impeller in units of less than 800 tons (cannot use an economizer)

Table 3-5   Advantages and Disadvantages of Medium- and High-Pressure Chillers

Multiple Impeller Design

Some centrifugal chillers are designed with more than one impeller. With this design, each impeller compresses the refrigerant gas in stages. Chillers with more than one impeller are referred to as multiple-stage compressors. Each impeller "throws" the gas and creates gas velocity, which provides the "lift," or the increase in pressure and temperature of the refrigerant as it is transferred from the evaporator to the condenser. The gas velocity may be high or low depending on the type of refrigerant used.

In effect, impellers change impeller rotation speed into pressure. To achieve higher pressure, required by some refrigerants, you need higher impeller rotation speed and gas velocity.

Systems with multiple-stage compressors can utilize a chiller economizer, a device that directs "flash" (evaporated) refrigerant gas that occurs as the pressure is dropped between the condenser and evaporator to an intermediate impeller stage (or stages). The lower pressure "flash" gas is useless at this point because it has gone through a phase change and is diverted through an intermediate stage to increase the pressure and temperature. This significantly improves chiller efficiency. Using an economizer is not possible with single-stage chillers since all compression is done by one impeller.

The CVHE CenTraVac® Chiller
Three-Stage Operating Cycle

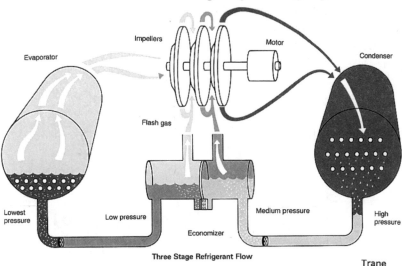

Impellers

Motor

Evaporator

Condenser

Flash gas

Lowest
pressure

Low pressure

Medium pressure

High
pressure

Economizer

Three Stage Refrigerant Flow

Trane

Drive Train Efficiency

The drive of an electric chiller consists of a motor that converts three-phase electric current into the energy of a rotating compressor shaft. A direct-drive, 60-Hertz, two-pole electric motor has a maximum speed of 3600 RPM, which is as fast as a standard two-pole motor can turn operating on 60 cycle alternating current.

Low-pressure impeller design works well with this direct-drive design since the pressure "lift" required (the difference in pressure between the evaporator and condenser) is only about 15 psi. The cooling effect is achieved with a relative low impeller gas velocity compared to medium-pressure chillers.

To achieve the compression and faster impeller rotation speeds needed for medium- and high-pressure chillers, gears or a transmission are necessary. A gear drive makes possible the higher RPM and gas velocity needed by medium- and high-pressure chillers where the lift is greater than 90 psi.

Gears or a transmission have an associated efficiency loss of one to two percent when the extra bearing losses are accounted for. Also, they have more parts which makes them somewhat less reliable. The added cost, maintenance, and energy use for medium- and high-pressure chillers are drawbacks that manufacturers attempt to overcome by making medium- and high-pressure systems smaller in size.

The diagram below demonstrates a direct drive and a hi-speed gear drive.

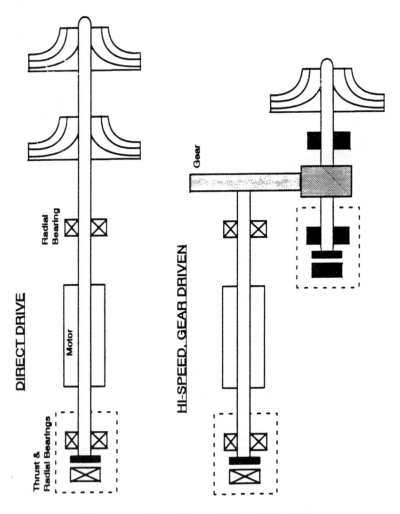

Figure 3-12   Direct Drive vs. Hi-Speed Gear Drive.

"A CASE IN POINT"

Recently, at a California university, the evaluation to replace a 200-ton centrifugal R-11 chiller with a 160-ton chiller lead to an interesting conclusion. The new replacement chiller required less cooling tons because lighting and other heat loads were reduced. However, a significant portion of the operation is now below 40% load. This means that low-load performance, which reduces efficiency has become a concern. The following is extracted from a report on the system evaluation.

Objectives

At the time of the study, the objectives of management were to:

1) Select a long-term, zero-chlorine refrigerant.

2) Reduce operating cost and achieve a reasonable return on investment.

3) As a technology-oriented university, implement a leading-edge showcase chiller system.

4) Gain support from industry manufacturers for demonstration.

Equipment Availability

An extensive review of current chiller technologies was performed to identify which chiller and refrigerant would best meet the objectives outlined above for the building.

R-22

An R-22 (HCFC) chiller was considered; however, R-22 is an interim refrigerant that contains chlorine and will eventually be phased out. R-22 did not meet the goal of a long-term refrigerant choice, and therefore, this option was rejected.

R-123

A centrifugal chiller that uses R-123, which is also an interim HCFC refrigerant, was considered because of its exceptional efficiency. It was discovered that while centrifugal chillers are very efficient in large sizes, a 160-ton chiller is small for a centrifugal chiller and was not as efficient as expected or as required. Because this chiller had only one compressor, part-load performance was not good and a single compressor did not provide a backup for system redundancy. It is not surprising that the highest efficiency machine was considerably more expensive relative to a chiller with average efficiency.

R-134a

An R-134a refrigerant, which is an HFC and does not contain chlorine, was considered for use in both a centrifugal chiller and a screw chiller. It was discovered that the centrifugal chiller met the cost criteria and also the long-term refrigerant criteria; however, the single centrifugal compressor did not provide for a backup, and efficiency was somewhat compromised because of the small tonnage. There also was some concern about high-speed gear reliability.

A single screw compressor could be utilized but also faced low-load problems and there was no backup compressor. There was one manufacturer that had dual hermetic compressors available. The screw chillers were said to work well with R-134a and the efficiencies were fairly good. However, like R-12, R-134a required larger machinery and heat exchanger surfaces and consequently had a price increase over a comparable R-22 screw chiller. Further investigation into the R-134a machinery led to the conclusion that it was not the optimum refrigerant for a chiller of this 160-ton size.

R-407C

The conclusion regarding R-134a led to the additional investigation of emerging system designs and refrigerants. There is a great international effort to develop alternative refrigerants for R-22 which has always been the preferred refrigerant in screw machinery. R-407C, which is a blended refrigerant of three components, has surfaced to the top as a good contender for an R-22 replacement.

At least two refrigerant manufacturers are marketing R-407C at this time and a recent research project revealed that this refrigerant can have an equal, if not better, efficiency than R-22.

Proposed Design

Based on the technology and equipment search, coupled with the availability of R-407C, the building management chose to meet their goals through the design and installation of a chiller that represents a culmination of the cutting-edge technologies not yet available in the United States.

The design would incorporate the following elements:

- Parallel screw compressor system with R-407C refrigerant.

- Welded flat plate chiller and semi-welded flat plate condenser, or new design counterflow shell and tube.

- DX chiller with electronic expansion valves.

- Refrigerant micro charge design.

- Microprocessor control system.

- High-efficiency heat rejection (variable speed and approach control).

Objectives Were Met

As things would have it, this chiller was fabricated at a facility just a few miles from the campus. The Electric Power Research Institute, EPRI, has taken an interest in the project and has provided precision instrumentation in order to compare operating efficiencies to the R-22 alternative. A part-load efficiency of 0.4 kW/ton and full load of .65 is expected.

The initial project objectives were met with the following elements:

- R-407C refrigerant: an HFC zeotropic blend R-32/125/134a (23/25/52) with a temperature glide of 10 - 12° F.

- Counterflow design of the evaporator and condenser with special circuits and baffling to take advantage of the glide characteristic of the refrigerant.

- Four 35 HP screw compressors, paralleled with an unloading sequencer and intermediate port economizer. (Given the configuration of the four compressors, the system can continue operating in the event one compressor fails. The compressors were bought "off the shelf," had a relatively low cost, and can be replaced in a matter of a couple of hours.)

- Braised plate sub-cooler/economizer.

- Micro-processor control with graphic interface, floating suction and head pressure control with cooling tower variable speed control.

The chiller is presently installed and operating. Data collection will take place through the cooling season of 1996 and test results should be known shortly thereafter.

Figure 3-13 R-407C Chiller

CHILLER MANUFACTURERS

The chart on the following page lists the major chiller manufacturers and a sample of the range of products they offer.

CHAPTER SUMMARY

A complete exploration of air-conditioning equipment options, dictated by the phaseout of CFC and HCFC refrigerants considers three basic options:

- Preserve current equipment and CFC refrigerants and concentrate on refrigerant containment, more efficient purge systems, and diligent maintenance.

- Convert existing equipment to a non-CFC alternative refrigerant.

- Buy new equipment.

Making use of new alternative refrigerants will have a substantial impact on new chiller design and on conversion alternatives. Most chillers incorporate one of four types of compressors: reciprocating,

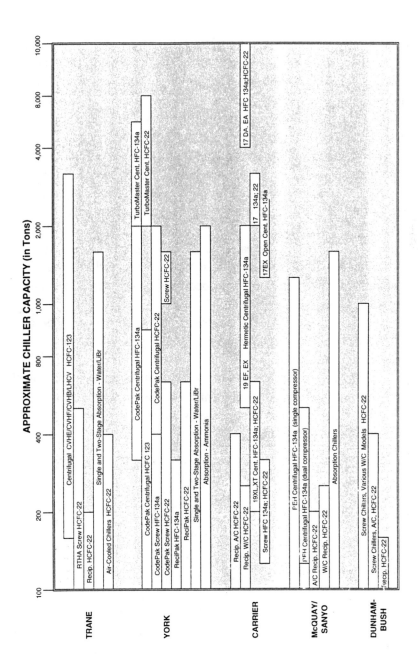

Table 3-6   Chiller Manufacturers and Products

screw, scroll, and centrifugal. Unfortunately, there aren't any "drop-in" refrigerant solutions for air-conditioning chiller equipment. New designs or conversions for each type of machine will need to accommodate the particular properties of the current and future alternative refrigerants, including compressor efficiency, operating pressures, impeller design and the number of impellers or stages, and heat exchanger design and size.

Conversion usually means the choice of refrigerant is limited by the type of compressor, size and construction of the chiller, and other factors. And once a conversion refrigerant is selected, converting a chiller to run with a substitute refrigerant usually requires some modification to the compressor or lubricant used.

New refrigerant properties, especially refrigerant blends that have "glide" characteristics, will more than likely dictate a change from the typical cross flow evaporator and condenser flow patterns to counter-flow designs.

In addition to these considerations, a refrigerant choice must also consider the availability and suitability of the new refrigerants to air-conditioning requirements.

In the meantime, refrigerant testing and analysis continue the search for efficient, safe, and environmentally friendly alternatives. Also, experiments continue for improvements in chiller system designs to make use of new refrigerants.

Chapter 4

Machinery Choices for Commercial Refrigeration Systems

In this chapter, we will explore the various commercial and industrial refrigeration machinery and system choices and alternative refrigerants you can consider as you put together your refrigerant management plan. This chapter concentrates on equipment and refrigerants used by supermarket and restaurant chains, and food-processing and cold-storage facilities.

INTRODUCTION

In addition to those aspects that are specific to refrigeration systems, the same factors that influence decisions regarding air-conditioning machinery choices also apply to commercial refrigeration systems. As in Chapter 3, this chapter discusses, in light of regulations and new refrigerants options available, the issues and options you must consider. Again, we will try to help you answer the question, What refrigeration equipment or refrigerant is best suited to your facility and applications?

The first part of this chapter introduces the three basic options you have regarding what to do with existing refrigeration equipment. Existing refrigeration equipment requires strict maintenance procedures and a guaranteed supply of refrigerant. Deciding to convert existing equipment or to purchase new equipment leads to considering the various refrigerant and machinery choices.

This chapter also will focus on the impact that different refrigerants have on commercial refrigeration machinery and system design. Since most commercial refrigeration systems are individually designed, each application has a unique set of issues: cost, availability, efficiency, and environmental impact.

This chapter will overview the two basic types of refrigeration systems: conventional (or one-on-one) and parallel (or multiplex) designs. We'll also explore how the choice of a particular refrigerant can have an effect on system design and components—compressors and their design; operating temperatures; operating pressures and their effect on relief and expansion valves and valve control applications; and other issues.

We'll also point out several system enhancements that can improve energy efficiency and provide demand savings (and help offset the cost of conversions and new purchases) including subcooling and floating head pressure.

A change in refrigerant may also mean a change in system lubricants or oils. We'll discuss this topic and also the primary refrigerant options for interim and long-term applications, and options for retrofitting existing systems and for new refrigeration installations.

There is one major case in point included in this chapter that describes in detail a refrigerant research project that tested and compared refrigerants in a test facility. This project simulated supermarket conditions with low- and medium-temperature refrigeration line-ups. An Impact Analysis of a study of over 80 typical supermarkets is summarized in a typical, "prototype" store. Another case study summarizes EPRI research.

OPTIONS FOR EXISTING REFRIGERATION EQUIPMENT

As you make decisions on how to react to new and future regulations for commercial refrigeration equipment, once again there are three basic options:

• Preserve existing equipment and CFC refrigerant.

• Convert existing equipment, or convert existing refrigerant, or both.

• Discard existing equipment and buy new equipment.

As with air-conditioning equipment, this is not a decision you make lightly since there will be long-term economic and environmental consequences.

Owners of supermarkets and commercial refrigeration are justifiably concerned about the capital investment required to comply with the phaseout of CFCs. Supermarket energy costs are high—on average, energy costs equal the net profit margin of about one percent. One-third to one-half of those energy costs pay for running refrigeration systems. Selecting the proper systems for a given application and taking advantage of refrigeration efficiency measures improve the potential for increasing that slim profit margin.

When selecting equipment, accessories, and controls for new or retrofitted commercial refrigeration systems, a system designer or contractor must consider many factors. For systems that are to be converted or retrofitted these factors include:

- Equipment condition, longevity, adaptability.
- Materials compatibility including lubricants, seals, motor windings.
- Code and permit limitations.

For new systems, the factors to consider include:

- The intended application—what kind of business or process is involved.
- The required refrigeration load and operating conditions.
- The refrigerant choices to be considered for the application (which impact cost, capacity, and power use).

All available equipment and refrigerant choices need to be considered. Also, there is a wide range of other factors including operating costs, any utility company assistance that might be available, total system energy efficiency, refrigerant availability, and any change in capacity you might need if there are plans to remodel the facility.

New refrigerants require machinery that is designed for the proper pressures, mass flows, and other characteristics. Many of the new refrigerants operate at substantially higher pressures, affecting not only compressors, but piping and vessels as well.

Unlike some air-conditioning chillers, most commercial refrigeration systems are custom designed or applied to match the required load.

The necessary components are combined—compressors, evaporators, condensers, and controls—to produce the desired capacity and operating characteristics. There will be a much greater range of possible solutions in terms of refrigerant and equipment options when compared with air-conditioning equipment. Also, keep in mind that manufacturers of refrigeration equipment tend to provide less advice on system configurations and conversions.

If, as a result your refrigerant management plan analysis, you have decided not to preserve some of your existing equipment and CFC refrigerant (as detailed in Option 1), the remaining options are to convert the existing equipment or refrigerant (Option 2), or to buy new equipment (Option 3). (The decision to maintain a particular CFC system or to convert to a non-CFC refrigerant is also discussed in Chapter 5.)

For Option 2, you could retrofit the system to use short-term (or interim) refrigerants (such as R-22, and blends that contain HCFCs and therefore chlorine), or retrofit the system to use long-term refrigerant alternatives that do not contain chlorine (such as pure HFCs or blends that contain HFCs.)

If you decide to buy new refrigeration equipment to replace existing equipment (Option 3), two viable options are to install new equipment that uses R-22, or that uses a long-term HFC refrigerant.

Option 1: Preserve Existing Equipment and CFC Refrigerant

Existing refrigeration systems that use CFCs can operate indefinitely if there is refrigerant to use in them. That is, you can continue to operate these systems if there is no loss of refrigerant, or if refrigerant has been stockpiled or is available from other systems as they are upgraded. Your refrigerant plan may include reclaiming and reusing refrigerant from retired systems.

The main goals of this option are to eliminate refrigerant loss and to avoid the cost of conversion or replacement for as long as possible.

An important aspect of preserving existing equipment and the CFC refrigerants they use is that many supermarket chains have short remodel cycles. That is, supermarket refrigeration systems, unlike chiller systems with around a 30-year life, are typically remodeled or replaced every seven to ten years.

Be aware there are potentially some risks with this option. CFC refrigerants on the open market will be expensive. If CFCs are used in existing systems you want to maintain, these refrigerants must be available either from existing stockpiles or from other systems as they are upgraded. And if CFCs cannot be obtained, you may have unplanned down-time or may need to bear the high costs of an unplanned system conversion.

If you determine that your current equipment has some life left and warrants keeping for the time being, you need to act on two key ideas to preserve your valuable supply of CFC refrigerant:

- Find and fix existing leaks.
- Reduce the potential for leaks.

Note that the discussion that follows is good advice for all refrigerants and systems both from an environmental and financial standpoint. Preventing leaks, and sound maintenance and servicing practices are especially important for systems that continue to use CFCs primarily because of their increasing scarcity and that the cost of CFCs have increased more than ten-fold.

Find and Repair Existing Leaks

To reduce refrigerant loss and to avoid the cost of conversion or replacement for as long as possible, you need to find as many leaks as possible in your refrigeration systems and fix them. This requires that you perform extensive leak testing and perform all necessary repairs to get the system "tight" and to take steps to contain your refrigerants and reduce refrigerant loss.

All refrigeration systems are susceptible to leaking refrigerants. If you knowingly allow refrigerants to leak, it is not only illegal but expensive.

Commercial refrigeration systems are perhaps more susceptible to leaks than air-conditioning equipment, specifically because of the hundreds of feet of pipe these systems typically use to carry refrigerant. Also, refrigeration systems contain considerably more refrigerant when compared to air conditioning—from four to 10 times the amount per quantity of cooling.

Reducing leaks is not only an issue for older systems, but for newer ones as well. The amount of refrigerant charge required for new facilities is as much as for older systems. While progress has been made in the area of reducing leak rates, owners and contractors report that new systems are far from leak-free and, in fact, new HFC installations have the same leak rates as the previous CFC or R-22 installations. (After all, the underlying system technology has not changed.) Once again, this results in increased operating costs because these refrigerants are expensive.

One simple indication that a system is leaking is that you keep needing to add refrigerant. A look at system recharge records or "leakage" logs can give you an indication. (This is discussed as a function of system maintenance below.)

At some sites with conventional one-compressor-to-one-line-up systems, there may be less refrigerant in each conventional system, but multiple separate systems in each store can offer more opportunities for smaller leaks. Conventional systems signal a leak by "flashing" or the presence of bubbles in the liquid line sight glass, which indicate the refrigerant level is low.

Larger parallel systems can have a much larger single-leak incident (larger leak, more refrigerant lost at one time). Newer installations have begun to use liquid level monitors (a sight glass gauge on the receiver) and refrigerant detectors to reduce the time necessary to detect a leak. (The receiver, which holds refrigerant and allows for expansion and contraction of the fluids, is a good location to monitor the refrigerant level.) Freezer warehouses typically have large R-502 systems which have a potential for substantial leakage and can require a lot of refrigerant to recharge.

Reduce the Potential for Leaks

There are many ways to minimize refrigerant loss and reduce the potential for leaks, not only with older existing refrigeration equipment, but with converted and new equipment as well. In this section we'll discuss how you can:

- Improve equipment maintenance and keep refrigerant use records.
- Improve refrigerant containment by performing leak testing, and by installing detection systems where feasible.
- Follow sound servicing procedures and practices, including refrigerant removal and system evacuation.

Improve Equipment Maintenance

One way to reduce leaks out of a refrigeration system is proper equipment maintenance. This primarily involves the simple task of tightening threaded connections on a regular basis.

During routine maintenance, all systems should be visually inspected to identify possible leaks or low refrigerant level. Also, maintenance crews can keep an eye open for fittings that need replacing or tightening, for capillary tubes that need replacing, and other system components that show excessive wear.

An important part of maintenance, which will also help with containment efforts, is to keep good records on all of your refrigeration equipment.

Individual logs or inventory sheets, for all units, should include information on the type of refrigerant, the amount used, capacity in tons, and charge size. Also important are recharging or "leakage" logs that include how much refrigerant is added to each system. As a way to detect leaks, keep a log of receiver refrigerant levels, if present.

An assessment of the equipment's leakage rate can help you determine how much refrigerant you will need each year to keep a particular piece of equipment operating. Keep these figures up-to-date as equipment is retrofitted, converted, or replaced.

Improve Leak Detection Efforts

Another important way to minimize refrigerant loss and reduce the potential for leaks is to improve your leak detection efforts.

Leaks from equipment and pipes can be detected using several methods, but the most efficient is to use portable, hand-held electronic detectors to pinpoint a leak at some specific system location. Also, concentrations of refrigerant vapor can be detected in an area or room using stationary monitors. The leak can then be isolated using a pinpointing device. Where possible, install detectors at all existing sites that have microprocessor control systems. Additional sensors can be used to measure reduced oxygen levels.

Note that leak detection equipment is necessary to comply with safety rules and safety recommendations. Using sensors to monitor leaks meets the recommendations of ASHRAE Standards 15-1992 and 15-

1994 and requirements of the Uniform Mechanical Code (UMC). (You must know and conform to all applicable codes.)

Too often, systems are not checked for leaks after a start-up following a refrigerant recharge, service, repair, etc. There are many factors that can cause a leak after start-up even if the system was "tight" prior to evacuation. After the system has run for a while and the high side components are hot and the low side components cold, a thorough leak test should be performed with an electronic leak detector.

(More about leak detection systems is found in Chapter 5 under "Implementing a Leak-Detection System." This section is part of the refrigerant management plan and discusses types of detection systems, detector sensitivity and selectivity, and tips for selecting an area monitor.)

Follow Sound Servicing Practices

Sound servicing practices can reduce the potential for leaks and improve containment in refrigeration systems. There are regulations regarding the legal way to service refrigeration equipment. Not only will proper servicing techniques reduce refrigerant losses, it will, along with proper documentation and good operating practice, avoid the potential for legal problems. In Chapter 5, these issues are discussed in greater detail.

Three important aspects of servicing are:

• Techniques for partial or complete removal of the refrigerant charge.

• System evacuation and dehydration methods.

• Pressure testing.

Refrigerant Removal

Refrigerant is removed from a system any time repairs, service, or modifications are needed, and also when a system is converted to use a different refrigerant.

When servicing an existing refrigeration system usually only a partial removal of the refrigerant is required—as is the case when a single component is replaced or serviced, or to repair a leak in a refrigerant line.

A system modification that reduces the potential for leaks when making repairs is the installation of isolation valves along the refrigerant line pipes. With these in place, you would only need to remove the refrigerant from one section of the system, only at the location the repair needs to be made. On larger systems, a permanently installed recovery system and storage tank allows faster and safer removal of refrigerant for servicing or system changes.

The typical procedure for replacing a component, as when you're replacing a burned-out compressor, is to isolate the compressor by closing valves in the refrigerant line on both sides of the compressor, remove refrigerant only from the isolated area, replace the compressor, evacuate the air that was admitted to the lines ("pull a vacuum"), and recharge the system with refrigerant.

If a leak needs to be fixed by soldering a pipe, and isolation valves are not installed, the total refrigeration charge from the system needs to be removed. If you don't remove the charge from the affected area, there is too much pressure for the solder to take (the solder is blown out of the opening). This means you need to remove all (or certainly most) of the refrigerant to reduce the pressure.

System Evacuation and Dehydration

There are several ways air can get into the refrigeration lines of a refrigeration system. If a repair or modification to the system requires that refrigerant lines be cut, and the lines are exposed to the air, air and moisture are admitted into the system.

The air that gets in, which includes non-refrigerant (non-condensable) gases (such as nitrogen, oxygen, etc.), needs to be removed to assure the refrigerant and system operate correctly. This is done by evacuating the system to as near a vacuum as possible. (Chillers perform continuous, on-going evacuation called "purging.")

Again, as with refrigerant removal, isolation valves permit evacuation of one section of the system. In this case, very little moisture will be admitted and only a partial evacuation is necessary.

However, a total system evacuation is required at other times. With a full system evacuation usually everything is the system is removed, including all of the refrigerant and air.

A full system evacuation is an important procedure that has often been overlooked or poorly executed usually due to time pressures, a lack of proper equipment, or a lack of knowledge.

A complete system evacuation is required:

- For a newly built system as part of the initial post-construction start-up routine.
- When the refrigerant in the system is changed.
- After a major system reconstruction, such as a change in case locations or pipe routing.

In these instances, a full system evacuation is part of "proofing" the system and is important because:

- A system that maintains a "deep" vacuum is likely to be leak-free. A system that doesn't hold a vacuum probably still has multiple leaks.
- New refrigerants and oils are less tolerant of moisture in the system and operation is unpredictable at best.

Most modern systems contain a number of regulators, check valves, three-way valves, and the like. Frequently, a complete system evacuation is not achieved because too few connections are made to the system. For instance, even in the direction of the flow, a check valve is effectively closed at the low pressures required during evacuation. To obtain a complete evacuation, the vacuum pump must be connected to both sides of the check valve. When connecting the vacuum pump, look at all parts of the system and determine whether there is a clear path from each point to the vacuum pump without passing through any pilot-operated regulators, float valves, or check valves that cannot be manually opened.

Another technique is to install vacuum sensors at multiple points on the system. Connecting the sensor at the vacuum pump—the typical method—does not always show the real vacuum in the system.

Another way to ensure there are no leaks after a retrofit, system service, or any other time the system is opened for repair, is to follow proper pressure testing practice.

A complete "system check" requires an initial evacuation, a pressure test, and a second evacuation to verify the system is holding a vacuum and is not leaking.

Some general guidelines are detailed on the following page in the boxes "General Guidelines for Pressure Testing" and "General Guidelines for System Evacuation." Always refer to RSES bulletins and the system manufacturer's recommendations for proper and complete evacuation and pressure-testing procedures.

General Guidelines for Pressure Testing

1. Use only dry nitrogen for pressurization. Never use compressed air.

2. Pressure test the system in steps with increasing pressure. (The number of steps depends on the size and complexity of the system.)

3. If the system has no immediately observable leaks, bring the system to one half of the design pressure.

4. Test the system for leaks using a mixture of liquid soap and water.

5. Repair any leaks found and repressurize the system with nitrogen to design (or test) pressure. (Test pressure could be higher than design pressure.)

6. Test the system for leaks again with soapy water.

7. Repair any leaks. Re-pressurize with refrigerant to 5-10 psig pressure and then raise pressure to at least 150 psig with nitrogen.

8. Leak check the system with an electronic leak detector suitable for the refrigerant being used. Repair any leaks found and repressurize if necessary until the system will hold pressure without dropping for 24 hours. (Note that if ambient temperatures drop, the pressure will drop too. This possibility must be accounted for.)

General Guidelines for System Evacuation

1. Connect the vacuum pump to the system at multiple locations to allow complete evacuation. Use a two-valve vacuum pump "tree" to allow connecting the vacuum gauge to either the pump or the system without removing any fittings.

2. Use 3/8" tubing or larger. For effective evacuation approximately one 3/8" connection should be made for each CFM of vacuum pump capacity.

3. Use an electronic "micron-type" vacuum gauge at the pump and at least one system connection point.

4. Evacuate the system to 1,000 microns or lower. The reading should hold constant for at least one-half hour with the pump valve off.

5. Break the vacuum with dry nitrogen.

6. Evacuate again to 1,000 microns or lower.

7. Break the vacuum with refrigerant.

8. Install filter and drier cartridges.

9. Evacuate the system to 500 microns or less. For low-temperature systems or evacuation during cold weather, lower vacuums should be used.

10. Break the vacuum with refrigerant. All blends should be charged in liquid phase only to avoid fractionation.

NOTE: Always refer to RSES bulletins and the system manufacturer's recommendations for evacuation and pressure-testing procedures.

Option 2: Convert Existing Equipment

If you have decided that your existing CFC refrigerant systems cannot be preserved, the second option is to convert them to use non-CFC refrigerants.

There are several key issues that need to be considered when converting existing systems to next-generation refrigerants:

- Conversion economics.
- System performance.
- Lubricant compatibility and stability.
- Materials or elastomer compatibility.
- Hardware changes.
- Disposal of the old refrigerant and lubricant.

All of these issues, and a sound strategy for proper system modifications, need careful consideration. Otherwise a system conversion could lead to a significant reduction in system life, system efficiency, and refrigeration capacity.

Conversion Economics

Converting a system to a non-CFC refrigerant is usually comparatively more expensive than preserving and maintaining your existing equipment and refrigerant. But if current systems are old and not functioning efficiently, or are experiencing high leakage rates, or if a store remodel dictates, it makes sense to convert. On the other hand, if a system is "tight" and is not experiencing any leaks or has a very low leak rate and refrigerant is available, it may make sense to keep it operating as is. But systems that have high leak rates are prime candidates for conversion simply because the cost of recharging them with CFC refrigerants may cost, over the long run, more than the cost of a conversion.

As with air-conditioning systems, the decision to convert commercial refrigeration systems to non-CFC refrigerants is often based on a corporate response to environmental regulations, a response to perceived or real shortage of CFC refrigerants, or if there has been a total loss of CFC refrigerant due to a catastrophic system failure.

However, with commercial refrigeration systems, the decision for some is based, in large part, on short-term economics. Since supermarkets and other commercial refrigeration users operate on very thin profit margins, some chains cannot accommodate long-term payback periods. For others, payback time frames are not an issue and conversions and store remodel expenses are just a cost of doing business. For different companies, you'll find a variety of economic considerations.

System Performance

In spite of the fact that a converted refrigeration system may lose cooling capacity and may use more energy, using a particular refrigerant in specific system configurations may have benefits. A conversion plan should explore capacity and energy use issues and may identify offsetting energy-saving opportunities, such as floating head pressure, subcooling, lighting conversions, control systems or other add-ons that can reduce the overall cooling load. An investment in system conversions may be justified based on energy use, equipment condition, and long-term store viability.

Other Conversion Issues

Other conversion issues relate to the physical condition and components of the existing system. Converting an existing system to use a different refrigerant brings up the issues of lubricant compatibility and stability (will the system and new refrigerant accommodate mixtures of mineral oil and synthetic lubricants?) the physical materials and elastomer components of the system (will they be compatible?) and possible hardware changes a conversion will require. Also, the old refrigerant and lubricant must be disposed of properly. You may have another system that could make use of the refrigerant that is being replaced (after it is reclaimed or recycled). In any case, the old refrigerant and lubricant should not be vented into the atmosphere.

Basic Conversion Options

There are two basic conversion options: retrofit the system to use short-term (or interim) refrigerants (such as R-22, and blends that contain HCFCs), or retrofit the system to use long-term refrigerant alternatives that do not contain chlorine (such as pure HFCs or blends that contain HFCs.) In general, the age and current condition of the system in addition to the system's application will strongly influence this decision.

For some situations, R-22 (or blends that contain R-22) may be the best choice as a "long-term interim refrigerant" because of the short remodel cycles (typically every seven to ten years) that many supermarket chains have. Also, R-22 may provide lower costs for systems that have high leak rates and there is always a possibility that HFCs may have other problems not yet discovered.

Given these two general conversion options for commercial refrigeration systems, there are many possible variables in terms of the refrigerant used and the design of the system. These retrofit options are discussed in detail later in this chapter after we explore the main types of refrigeration systems and design considerations.

Also, issues regarding retrofit options for existing R-12 medium-temperature systems, and for existing R-502 low- and medium-temperature systems are discussed later in this chapter.

Option 3: Buy New Equipment

The third option is to plan for purchasing new refrigeration equipment and the additional system components you may need. If the existing equipment is at or near the end of its useful life, chances are the system is not operating efficiently. So, for stores or facilities with good future potential, instead of spending money on conversion or improved containment of the old systems, investing in a new system may be the best alternative.

The best candidates for new refrigeration systems have the characteristics described in the table on the following page.

In some cases, remodeling the entire store sometimes justifies buying new machinery instead of using existing equipment. Any anticipated new system configuration should include flexibility for future expansion and should incorporate features that allow easy re-configuration.

The new refrigerant choice of course will impact the machinery choice, system design, efficiency, and type of compressor. And once again, the specific system design and refrigerants used will depend on the applications, temperature requirements, and other factors.

If you're going to install new commercial refrigeration equipment, a parallel system design (described later) that uses one of the new, long-term refrigerants offers the best overall economic and environmental choice. This type of system may require more up-front expense, but because it is much more energy efficient, impact studies have shown the energy savings can pay for the new equipment in five to eight years.

The Best Candidates for New Refrigeration Systems	
Store is expected to stay in service for 15 years or more	In particular, any store that is anticipating remodeling would be a good candidate. A new system can be sized to allow flexibility for reconfiguration for future loads.
Relatively large store	Energy savings economics will be better in a larger store than in a small store.
Refrigeration equipment is fully depreciated	If the equipment is fully depreciated, you will avoid writing off existing assets.
Condenser or other equipment is in need of replacement	Any equipment that would otherwise need replacement improves the economics of the decision. It's better to contribute to a new system and reap a "cost avoidance"
High CFC leak rate and/or refrigerant charge	The CFCs in a system with a large refrigerant charge can be captured and used in existing systems in other stores.

Table 4-1   The Best Candidates for New Refrigeration Systems

(Note, as an example, an oversized condenser contributes to very high efficiency and would not be an unused investment.)

For a typical 30,000 to 40,000 square foot store, the economics of upgrading to a new system are in the range shown in the table below.

Typical New Refrigeration System Costs	
Implementation cost:	$220,000 to $260,000
Annual savings:	$45,000 to $50,000
Simple payback:	4.5 to 6.0 years

Table 4-2   Typical New System Costs

The rest of this chapter discusses the impact of new refrigerants on commercial refrigeration system design. Specific, detailed options for new medium- and low-temperature refrigeration equipment are discussed later in this chapter.

NEW REFRIGERANT IMPACT ON THE DESIGN OF REFRIGERATION EQUIPMENT

Utilizing new refrigerants will have a substantial impact on the design of equipment used in supermarket refrigeration systems that provide cooling for a store's refrigerated display cases and walk-in boxes.

To provide a context for later use, we'll begin with a very brief overview of the basic components and operations of two basic refrigeration systems: the conventional or one-on-one system, and the parallel or multiplex system. Then, we'll discuss compressor design options and how using new refrigerants will affect system design pressures and piping and valve components. Also, we'll discuss two design implementations that can improve system efficiency: subcoolers and floating head pressure.

Keep in mind that there are energy-consuming parts of refrigeration systems found in supermarkets not discussed in this section but are important to consider in terms of total system efficiency. These include case lights, circulating fans, "anti-sweat" strips on freezer case edges, defrosters, and others. Also, another important consideration is that high levels of humidity in a store can increase the load on refrigeration systems and can decrease display case efficiency by contributing to the formation of frost on evaporator coils.

Finally, the last topic of this section covers the use of lubricants with different refrigerants in refrigeration systems.

Types of Refrigeration Systems

There are two basic types of commercial refrigeration systems:

- "Conventional" or one-on-one system—this type has one display case line-up to one compressor.

- Parallel or multiplex systems—this type has several display case line-ups (or boxes) cooled by several compressors connected to a common "manifold."

A Conventional Refrigeration System

All supermarkets have a mix of low-, medium- or high-temperature display cases to meet specific product refrigeration requirements. Some supermarkets use open-type display cases and other chains may use glass-door cases for much of their frozen food products.

A very common refrigeration system configuration found in stores is the conventional (one-on-one) type.

Each conventional refrigeration system compressor has its own condenser/receiver, and each compressor serves a single line-up or several small display cases with similar saturated suction temperatures. Each compressor also has its own dedicated suction, discharge, and liquid lines. Conventional systems nearly always use CFC refrigerants and many are past their useful life. To provide refrigeration, the compressor in a conventional single-compressor system either operates at 100% capacity, or is off.

Figure 4-1 illustrates the typical conventional refrigeration systems found in many stores.

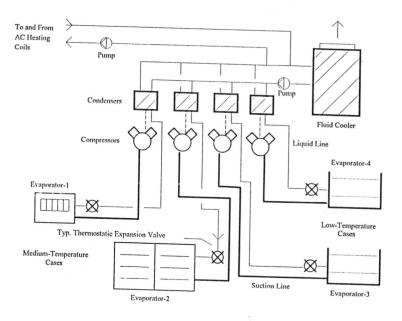

Figure 4-1 Conventional Refrigeration System

The most common method used for the refrigeration process is the vapor compression system, also called the simple compression cycle. The main components of the conventional refrigeration system are the compressor, condenser, expansion valve, and evaporator.

In the simple compression refrigeration cycle, the system maintains a low pressure at the evaporator, and a high pressure at the condenser. The refrigerant acts as a transportation medium to move heat from the evaporator to the condenser where the heat is given off, either to the ambient air, or in a water-cooled system to the cooling water. A change of state from liquid to vapor and back to liquid allows the refrigerant to absorb and discharge heat.

In more detail, high-pressure liquid refrigerant is fed from the condenser, through the liquid line, and through a filter-drier, which extracts water or moisture from the refrigerant.

The refrigerant continues on through the metering device which separates the high-pressure side of the system from the low-pressure evaporator. There are several types of pressure control devices. A typical type is the thermostatic expansion valve. The thermostatic expansion valve controls the feed of liquid refrigerant to the evaporator, and reduces the pressure of the refrigerant. This reduction of pressure causes it to boil or vaporize until the refrigerant is at the "saturation" temperature corresponding to its pressure.

As the low-temperature refrigerant passes through the evaporator coil, heat flows through the walls of the evaporator tubing to the refrigerant, causing the boiling action to continue until the refrigerant is completely vaporized.

The expansion valve regulates the flow through the evaporator as necessary to maintain a preset temperature difference or "superheat" between the evaporating refrigerant and the vapor leaving the evaporator.

The refrigerant vapor leaving the evaporator travels through the suction line to the compressor inlet. The compressor takes the low-pressure vapor and compresses it, increasing both the pressure and temperature. The hot, high-pressure gas is forced out the compressor discharge valve and to the condenser.

As the high pressure gas passes through the condenser, it is cooled by some external means. On air-cooled systems a fan and fin-type condenser surface is usually used. On water-cooled systems, a refrigerant-to-water heat exchanger is usually used. As the temperature of the refrigerant vapor reaches the saturation temperature corresponding to the high pressure in the condenser, the vapor condenses into a liquid and flows back to the receiver to repeat the cycle.

Often a single multi-circuit evaporative condenser (an evaporative condenser that serves multiple cooling circuits) or one closed-loop fluid cooler is used. In the latter instance, each conventional system has a water-cooled condenser connected by a common closed water loop to the fluid cooler. Occasionally this closed water-loop is used for store heating by means of a water coil in a central air handler.

Parallel or Multiplex Refrigeration System

Although many large supermarket chains continue to use conventional systems with one compressor dedicated to one case line-up, for new construction or for retrofits, parallel or multiplex refrigeration systems have become the systems of choice for most supermarkets over the last fifteen years.

This type of system has several display case line-ups (or boxes) that are cooled by several compressors connected to a common "manifold." This more complex method is quite efficient and is consistent with the best current technology, and typically employs high-efficiency heat rejection, and ambient or mechanical subcooling.

A parallel or multiplex system operates basically the same way as the conventional type of system except this type uses multiple compressors piped to common suction and discharge manifolds. It consists of three or four compressors sized so that operating all of the compressors at the same time can provide enough capacity for the required refrigeration load. For example, a system with three compressors of the same size may have four capacity steps: 100%, 67%, 33%, and all off (0%). This configuration provides substantial flexibility for varying system demand.

With a parallel or multiplex system, several enhancements can improve energy and demand savings: floating head pressure, hot gas defrost, and heat reclaim. (Thermal energy storage (TES) is another future energy-efficiency feature.) Research (by EPRI and others) has found that floating head pressure, hot gas defrost, and heat reclaim produce the most energy efficiency benefits. Also, by closely matching the display case temperatures to those of the compressors, and by using specialized HVAC

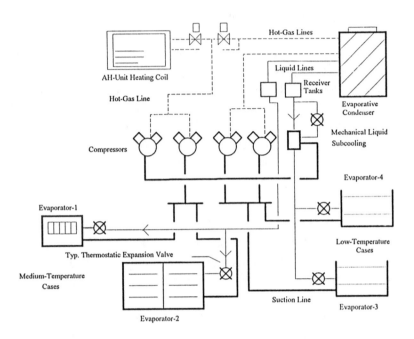

Figure 4-2 Parallel Refrigeration System

systems designed for reducing humidity, energy use is also improved. These features should be considered with any new system.

Figure 4-2 illustrates a basic high-efficiency parallel system.

Compressor Design

In the conventional system configuration, there are typically two types of compressors, categorized in terms of the temperature they maintain: medium-temperature and low-temperature compressors.

Medium-temperature compressors are used in systems that keep refrigerated product at above-freezing temperatures (see Table 4-3). Medium-temperature cases usually contain dairy products, beverages, meats, fruits, vegetables, and other non-frozen perishable products. Low-temperature compressors are used to preserve frozen foods. The average temperature of a frozen food case is 0 degrees F. The table also indicates the typical refrigerant Saturated Suction Temperature (SST) range for these compressors (the temperature of the refrigerant leaving the evaporator).

Low-temperature compressors tend to have less load variation than medium-temperature units. (Most frozen food cases are enclosed so temperature fluctuations are minimized.)

Typical Refrigeration Temperatures		
	Medium-Temp. Compressors	Low-Temp. Compressors
Product Temperature	30° to 42° F	0° to -20° F
Refrigerant Saturated Suction Temperature (SST)	5° to 25° F	-15° to -40° F

Table 4-3   Typical Refrigeration Temperatures

There are several issues regarding the type of compressor used in both retrofit and new systems. Some older compressors can be used in retrofit situations using new refrigerants. On the other hand, other existing compressors cannot handle the pressures needed for operation. In this case, the compressor must be changed to one that can accommodate the pressure requirement.

In the case where a system is converted to a refrigerant with the same or similar operating pressures as the original refrigerant, the compressor might be used in a retrofit situation. As examples, R-134a has approximately the same operating pressure as R-12; R-404A, and R-507 have approximately the same operating pressure as R-502; and R-407C approximately the same operating pressure as R-22. In these cases, the existing compressor and motor combinations might be compatible in retrofit systems, which would allow the use of previous compressor displacements and motor combinations.

The refrigeration industry basically uses the following types of compressors:

• Reciprocating compressors (hermetic or semi-hermetic; single-stage or two-stage).
• Screw compressors.
• Scroll compressors.

Compressor Characteristics and Trends			
Type	Size Range	Characteristics	Market Trends
Hermetic Reciprocating	1/3 to 5 HP	Low cost, small size	The only self-contained model, used with small and distributed refrigeration systems.
Semi-hermetic Reciprocating	1/3 to 60 HP	Reliable, wide application range	Have now established about 95% of the market share.
Scroll	1/3 to 12 HP	Low cost, small size, quiet operation	Small market penetration.
Screw	15 to 80 HP*	Large capacities, simple operation	Increasing usage.
* Screw compressors exceed 1,000 HP in industrial versions			

Table 4-4   Chiller Compressor Characteristics and Trends

The table above describes these types of compressors. Each of the types listed may be used in either medium- or low-temperature applications.

Reciprocating Compressors

As described in Chapter 3, reciprocating compressors use positive displacement by means of pistons to compress the refrigerant. The pistons move inside cylinders to reduce the volume of the vapor in the compression chamber. A reciprocating compressor may have one or more pistons. The pistons are usually driven directly through a pin and connecting rod from a crankshaft that is turned by an electric motor.

Reciprocating compressors may be classified as hermetic or semi-hermetic. Hermetic compressor designs use an enclosure around the compressor and the electric motor drive. In this type of design, the motor is cooled by refrigerant that circulates around the motor windings. Semi-hermetic compressors, like the hermetic design, use refrigerant to cool the electric motor, but this type is accessible through bolted covers.

Reciprocating compressors are typically used with R-22 and R-502 and are the most common type. Reciprocating compressors can be used with all of the new refrigerants.

Single- and Two-stage Applications

Reciprocating compressors can achieve the required pressure and discharge temperatures in a single stage, or in two stages using more than one compressor. Also, the required pressure and discharge temperatures can be achieved using multiple stages in a single compressor (an internally compounded two-stage compressor).

For low-temperature single-stage applications, the compressor must work harder to achieve the correct compression ratio. As the compression ratio increases, the efficiency decreases and the heat generated by compression increases. The low end of the evaporating temperatures for single-stage compressors is limited. So, in order to decrease evaporating temperatures and increase operating efficiency at low temperatures, the compression can be done in two steps or stages.

A two-stage system (also called a two-stage booster system) uses two individual compressors (or two banks of compressors) to raise the refrigerant to condensing temperature in two steps.

The refrigerant is compressed some in the first stage and then compressed more in the second stage. In this configuration, using two compressors, the discharge of one compressor (the low- or booster-stage compressor) is pumped into the suction inlet of the second (the high-stage compressor).

This method reduces the discharge temperatures and the possibility of oil breakdown often found with a single-stage compressor. Both Copeland, Corp. and Carlyle Compressor Co. offer compressors rated for booster application.

The interim use of R-22 benefits from the use of two-stage reciprocating compressor systems, with two complete compression cycles, or at least requires compressors designed for liquid-injection cooling, or other methods of reducing discharge temperatures to avoid oil breakdown and shortened equipment life.

Liquid Injection

Liquid injection compressors are designed to be cooled by injecting liquid refrigerant in between stages. In this design, a controlled amount of saturated liquid refrigerant is applied through a meter into the compressor suction cavity or line to cool the suction gas.

The temperature of the refrigerant vapor leaving the first stage and entering the second stage may be high due to the heat of compression, which can result in overheating of the second stage cylinders and valves. Liquid refrigerant injection prevents compressor damage by cooling the second compressor and lowering the discharge temperatures to a safe level. The meter ensures injection only occurs when it is required to lower discharge temperatures.

Generally, systems modified to use liquid-injection two-stage compression have been more expensive to maintain. Two-stage systems contain more parts and more controls, and are less durable under unusual conditions. Also, compressors designed with liquid injection have reduced capacity during high ambient conditions—just when capacity is needed the most.

Care must be taken not to allow excessive mass flow through the booster (low-stage) compressor. Particularly for systems with gas defrost or large electric defrost systems, short periods of high suction pressures can cause operation outside expected conditions, which can result in broken valves and compressor failures. Field retrofits on two-stage systems include the addition of a "holdback" valve on the booster discharge—reducing compressor failures but increasing energy use.

Internally Compounded Two-Stage Compressors

With two-stage systems that use two compressors, it is sometimes difficult to maintain proper oil levels in the two crankcases. Another option for achieving two-stage compression is by using a single compressor with multiple interconnected cylinders. In these so-called internally compounded two-stage compressors, the pressure and discharge temperatures are achieved using multiple stages in a single compressor. For example, in a reciprocating compressor with six cylinders, cylinders 1 through 4 perform the low-stage compression by increasing suction pressures from the load to the first level, then the remaining two cylinders perform the high-stage compression, increasing the vapor temperature to normal discharge levels.

A two-stage compressor is designed so that suction gas is drawn directly into the low-stage cylinders and then discharged into the high-stage cylinders. Each stage of compression then is at a much lower compression ratio and the compressor efficiency is greatly increased.

Screw Compressors

As discussed in Chapter 3, screw (or helical rotary) compressors, like reciprocating units, are positive displacement machines. They have either two matched spiral-grooved rotors, (intermeshing "screws") or one rotor with a "gaterotor." As they turn, the volume of the refrigerant chamber between the screws is reduced, compressing the refrigerant. The screw compressor has nearly constant-flow performance and can provide high-pressure ratios.

Screw compressors have been used in industrial refrigeration since the 1970s and are now considered the standard design approach in that segment. Because of the complexity and close tolerances required to manufacture the screw rotors, smaller machines have been slow to enter the market.

In the past three or four years, two manufacturers have achieved significant results in the commercial refrigeration industry, notably with grocery chains located on the East Coast. Compressors manufactured by Bitzer are available in sizes down to 25 horsepower for R-22 and for alternative R-22 refrigerants. Carlyle Compressor Co. offers screw compressors down to 15 HP for R-22, R-134a, R-502 and R-404A.

These compressors have a number of important differences in comparison with reciprocating compressors. First, although screw compressors use a large amount of lubricating oil, there is no crankcase. The oil is supplied from the oil separator in the system high side. No crankcase means the system must include oil filters, oil coolers (depending on application), and additional safety controls.

To achieve capacity control with small screw compressors, it is common to use variable speed drives.

Scroll Compressors

Scroll compressors have achieved a large share of the residential air-conditioning marketplace, offering quiet, efficient operation and high reliability. Scroll compressors for air-conditioning applications are offered by Copeland from 1 1/2 HP to 13 HP. Refrigeration models are currently offered by Copeland up to 5 HP.

Scroll machines have begun to enter the commercial refrigeration industry as an alternative to hermetic reciprocating compressors in self-contained equipment. Hussmann Refrigeration Co. has used the scroll compressor exclusively in its Protocol™ system. This innovative system distributes the refrigeration compressors throughout a supermarket in a number of "cabinets" that comprise a complete water-cooled parallel refrigeration system, in order to reduce the overall refrigerant charge.

New Compressor Development

Compressor manufacturers have invested heavily to develop compressors that are compatible with the new HFC refrigerants and associated lubricants. Most of the efforts have been directed at lubrication of the compressor. In a refrigeration system, there is always some refrigerant contained in the lubricant—which itself circulates throughout the system. The chlorine content of CFC molecules actually *improved* the lubrication performance of mineral oils. HFCs and Polyol Ester lubricants (or POEs, which are discussed later in this chapter) have different characteristics, requiring a great deal of research to produce long bearing life.

R-134a was the first refrigerant that compressor manufacturers addressed, and all major manufacturers offer compressors and ratings for R-134a. The major manufacturers also now have compressors rated for R-404A and R-507. Many also have developed ratings for R-407C at air-conditioning or medium-temperature conditions, as a replacement for R-22.

A few manufacturers have developed compressors for R-410A and R-410B, which operate at much higher pressures than any previous common refrigerants. Copeland is currently developing scroll compressors for R-410A and R-410B. In addition to being physically constructed to withstand higher design pressures, these compressors have smaller displacement (and therefore larger motors) than previous compressors. So far, compressors for R-410A are limited to hermetic models. In the future, larger semi-hermetic or screw compressors may be developed to exploit the advantages of this refrigerant.

After developing compressor models suitable for HFC refrigerants, compressor manufacturers must perform extensive testing on their compressors with the new refrigerants in order to develop performance data and to obtain UL listing for each model. Millions of dollars have been spent in this effort and, needless to say, the manufacturers are not anxious to develop ratings for more refrigerants than necessary.

Currently, most compressor manufactures have published ratings for R-134a, R-404A and R-507 in applicable low- and medium-temperature application ranges.

R-407A and R-407B are used more extensively in Europe than the U.S., so manufacturers have published ratings and certifications in Europe.

Other Refrigeration System Design Considerations

New refrigerants will have an impact on the design of new and converted refrigeration systems. Some of the important considerations are:

- System design pressure.

- Piping design.

- Valve and control application including relief valves, expansion valves, and solenoid valves.

- Subcooling including ambient and mechanical subcooling.

- Floating head pressure.

- Code considerations for new systems.

System Design Pressure

Systems that are converted from CFC refrigerants to use HCFC or HFC alternatives will most likely need to accommodate differences in pressures that the refrigerants operate under. In addition to compressor issues mentioned earlier, the receiver tanks, piping, related accessories, and safety equipment such as the relief valves, must be suitable for any system converted to use a new refrigerant.

Typically, the required design pressure is based on both the refrigerant and the type of condenser used. Since some new refrigerants operate a higher temperatures, changing from air-cooled to evaporative condensing could be considered as an option for some converted systems rather than replacing the pipes and vessel, which would be required if air-cooled condensing is kept. Evaporative condensers are able to maintain a lower pressure which means the old pipes and vessels will function with new refrigerants.

There are other similar issues to consider for converted systems operating at higher pressures. Pressure switches, particularly high-pressure switches may need to be changed to match the requirements of new refrigerants. Application of solenoid valves and regulators for new refrigerants requires adjustments for the different pressures and mass flow rates.

Regulators, used primarily with refrigeration systems, must be adjusted to match refrigerant flow at new pressures and mass flow rates. Suction and discharge regulators may need to be changed, depending on their design pressure range and the pressure of the new refrigerant.

Acceptable system design pressures are dictated by the various codes and standards applicable in a given locale. Equipment owners and contractors should be aware of the local codes, as well as the model code guidelines and industry consensus guidelines, that apply in particular geographical locations.

Tubing Design

The charts in Figures 4-3 and 4-4 show the saturation pressures at 130° F for various refrigerants and the tubing design pressures for a range of tubing sizes for both Type L and Type K wall thicknesses. Type L is the thickness standard for commercial refrigeration tubing. Type K has been used only occasionally in the past.

In this example, an air-cooled system using R-402A (at about 375 psig) would require heavy-wall, Type K tubing on all discharge lines 1-5/8" and larger. Note that several of the higher-pressure refrigerants require Type K tubing when the size exceeds 1-3/8."

Piping Design Guidelines

For converted or new refrigeration systems, piping design should include every reasonable effort to reduce refrigerant leaks. Some specific guidelines are:

* Minimize threaded joints. Use only "sweat" type expansion valves, shut-off valves, etc.

* Use only long radius elbows and avoid 45-degree elbows to reduce fitting breakage.

* Install piping with adequate expansion loops or direction changes to allow for expansion and contraction. Typical commercial applications should have a five-foot double offset or a ten-foot single offset for every 80 feet of straight run.

* Use high-temperature brazing alloy only—do not use soft solder.

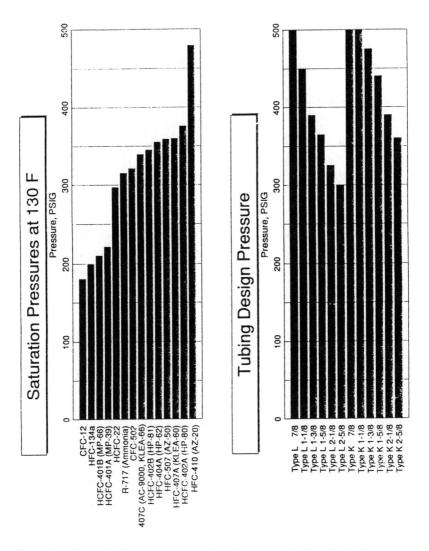

Figure 4-3 Saturation Pressures at 130° F   **Figure 4-4** Tubing Design Pressure

- Integrate logistically placed access and bypass valves into the piping system. Instead of small 1/4" access valves and self-closing "Schrader" valves, install 3/8" or larger angle valves at multiple locations. Consider how the system charge could be quickly removed from each portion of the system to allow servicing. This will facilitate refrigerant evacuation during system servicing and installation.

Valve and Flow Control Application

In addition to operating at varying pressures, different refrigerants require different flow rates to provide a ton of refrigeration. Valve manufacturers are rapidly developing ratings for the more common refrigerants used in converted and new systems.

Many adjustable valves in converted systems will probably need to be reset and general correction factors included for liquid line components such as solenoids. Solenoid valves have a maximum operating pressure differential (MOPD). If a system is changed to a higher-pressure refrigerant, the rating should be checked and the solenoid replaced if it is not adequate.

Also, discharge line and suction line components will require new components. Since refrigeration systems are typically custom-built, consult with a component manufacturer to determine the necessary adjustments and components.

Relief valves

Relief valves on commercial refrigeration systems are used to vent refrigerant that is pressurized beyond design parameters to the atmosphere in order to protect the vessels.

A number of factors have made relief valves an increasingly important subject for the refrigeration industry, including: the high cost of refrigerant and environmental considerations, use of new refrigerants with higher operating pressures, and new safety codes for mechanical refrigeration.

Relief valves are selected both in terms of pressure setting and mass flow capacity. Relief valves often re-seat tightly after an emergency relief occurs but they are not designed to operate repetitively and should be changed if they have been activated. After converting to a new refrigerant, don't assume that the existing relief valves will be adequate—a leaky or faulty relief valve can cause problems. When the refrigerant is changed, it's a good idea to change the relief valve to a properly sized valve, or better yet, a special three-way valve combined with two relief valves off a high-pressure receiver (see the following graphics).

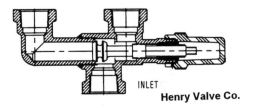

Henry Valve Co.

Figure 4-5   Three-way Dual Shut-off Valve

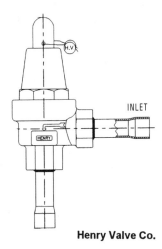

Henry Valve Co.

Figure 4-6   High/Low By-Pass Valves

One difficulty has been to know if and when a relief valve has functioned. One manufacturer offers a solution—a combination rupture disk
and pressure switch or transducer can be used to signal that a relief
valve has functioned. When a high-pressure event occurs, the disk ruptures and the relief valve opens. After the relief valve resets, pressure
remains on the pressure sensor, allowing an alarm to be generated.

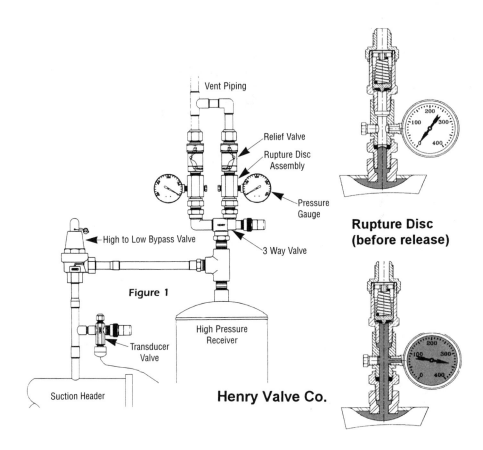

Vent Piping

Relief Valve

Rupture Disc Assembly

Pressure Gauge

High to Low Bypass Valve

3 Way Valve

Rupture Disc (before release)

Figure 1

High Pressure Receiver

Transducer Valve

Suction Header

Henry Valve Co.

Rupture Disc (after release)

A special type of relief valve, previously used mainly on shipboard applications, is now available in commercial sizes. This relief valve vents excessive high-discharge pressure to the system low side. This valve is used in addition to the atmospheric relief valve and does not replace it.

Assistance with proper selection of relief valves can be obtained from valve manufacturers, including Henry Valve Company, Superior Valve Company and others.

Expansion Valves

Expansion valves adjust refrigerant flow like a regulator. In converted systems, previously existing expansion valves are very often oversized. Older valves can also have considerable wear. Many of the long-term alternative refrigerants will require expansion valve adjustments. Careful attention must be paid particularly for refrigerants with glide. Since these refrigerants have different characteristics and behave differently, the expansion valve needs to be adjusted accordingly.

Subcooling

There are two types of subcooling: liquid-to-suction subcooling and mechanical subcooling.

Briefly, mechanical subcooling refers to the use of an extra compressor to partially cool hot liquid returning from the condenser before going to display fixtures, which reduces the work of the main compressor rack.

Liquid-to-suction subcooling takes cool suction gas from the evaporator to cool the liquid refrigerant line in a heat exchanger. The refrigerant temperature in the liquid line is reduced to below the saturated temperature, or the exact point at which the refrigerant changes from liquid to gas.

The purpose of subcooling is to provide more cooling capacity. Newer refrigerants don't carry heat as efficiently which means there is a loss of cooling capacity. If subcooling is added to the system, it will regain some or all of the lost capacity, and provide a more efficient system.

Subcoolers are used only with low-temperature systems. Both methods of subcooling have a significant impact when used with the new refrigerants. For HCFCs (such as R-22), subcooling provides up to 20% more cooling capacity. For HFCs (such as R-404A or R-507), subcooling provides up to 40% more cooling capacity.

Mechanical subcooling can be added to existing systems or designed into new ones. It can be applied in situations where more capacity is needed, or merely for reduced operating costs. The investment in subcooling can provide a quick payback period from reduced energy costs. Also, for new systems, compressors and control valves can be smaller.

Note that subcoolers should not be put on oversized systems where the cooling capacity is greater than the cooling load. In this case, subcoolers can short cycle the compressor, causing a repeated on/off, start/stop situation which can damage the compressor.

Liquid subcooling provides many benefits:

- It increases the effective capacity to a two-stage compressor. Since the liquid entering the evaporator has a much lower enthalpy content, it has greater heat absorption capability and therefore increases the number of BTUs each pound of refrigerant can absorb.

- Eliminates the flash gas within the system. This means a more efficient use of the evaporator surface area. Similarly, subcooling eliminates the need to increase evaporator surface area to increase capacity caused by larger loads.

- Decreases electricity use because of the reduced horsepower needed in a low-temperature system.

- Reduces maintenance.

- Gives better temperature control at the food cases.

- Reduces the time required to return the system to operating temperature after defrost ("pulldown" time). (In some cases, pulldown times are half what they are without the subcooler.)

- Can reduce first costs by producing equal capacity with smaller equipment.

- Allows for floating head pressure (described below). Floating head pressure is often only possible with subcooling. For parallel systems, liquid subcooling is the main means of accommodating floating head pressure.

Floating Head Pressure

Floating head pressure refers to increasing and decreasing or "floating" the compressor head or discharge pressure. The lower the head pressure, the less the compressor needs to work.

To take advantage of variable head pressure, subcooling control is required. With subcooling, condensing temperatures will fluctuate with changes in ambient conditions, and floating head pressure allows the compressor discharge pressure to be lowered as ambient temperature decreases. Subcooling and floating head pressure assures only liquid refrigerant will enter the expansion valve.

The key benefit of floating head pressure is that it saves on energy use. Floating head pressure reduces power consumption for all refrigeration compressors in systems with either an evaporative or air-cooled condenser. However, for parallel systems, the way heat reclaim or gas defrost are implemented may limit how far head pressure can float.

So, subcooling and floating head pressure may render greater benefits with some of the newer refrigerants.

Code Considerations

With the introduction of new refrigerants and more regulations, commercial refrigeration (new and converted systems) has come under greatly increased regulation from building and fire safety codes and officials. Often, commercial refrigeration had been considered a "fixture" rather than a part of a permanent construction and all but ignored by building inspectors. This is changing rapidly.

In the industrial sector, ammonia refrigeration has gone through dramatic changes over the last decade. Leakage has been greatly reduced—most modern plants have no or very little discernible ammonia smell. Safety training, technical certification and regulatory involvement have a very high priority with industrial refrigeration organizations.

With this experience, many regulatory officials are now turning their attention to commercial refrigeration and the new refrigerants. New requirements are being added to the building, mechanical, and fire codes at the same time old requirements are being more strictly enforced. The final outcome of many issues is not known at this time. For certain, though, knowledge of codes, standards, and other mechanical design requirements will be increasingly important.

Of particular importance are system design pressures as they relate to retrofit situations. These are dictated by the various codes and standards that are applicable in a given locale. In the past, most of these concerns were addressed by the original equipment manufacturer, who expected that the refrigerant in the system would be unlikely to change from the one used in its original design. Certainly, no designer anticipated the wide array of possible refrigerants and mixtures that now exists. The responsibility for a safe system lies with those involved.

There are many codes that regulate refrigeration system requirements including machinery room regulations, refrigerant detectors and shut-offs, alarm and exhaust fan evacuation, emergency shut-off and ventilation, requirement of self-contained breathing apparatus, and other aspects. Chapter 1 of this book describes some including ANSI/ASHRAE Standard 15-1994 Safety Standard, ASHRAE Standard 3. Others include standards set by the Underwriters' Laboratories (UL) and Uniform Mechanical Codes. As always, it's up to you to know about the regulations that affect you directly and to follow them closely.

Lubricants for Commercial Refrigeration

All refrigeration systems require specialized, high-quality oils to provide lubrication for the compressor. Some of the oil constantly circulates with the refrigerant through the system. High quality is important because one oil charge could be used essentially for the life of the system.

The particular lubricant used in a refrigeration system depends on the refrigerant used and on the application—because different refrigerants function at different temperatures and pressures, the lubricant must have compatible properties. Many of the new, alternative refrigerants require lubricants that are very different from conventional oils, many of which have been in use since the 1930s.

Lubricants must be able to mix appropriately with the refrigerant. This is referred to as the miscibility of the lubricant. As the oil circulates with the refrigerant, it is present in the evaporator and condenser of the system. If the two fluids mix correctly, the oil is carried back with the refrigerant to the compressor reservoir. If the refrigerant and oil don't mix correctly, operational problems can occur because the oil can accumulate in the condenser and evaporator. Unmixed oil in the evaporator can reduce heat transfer and constrict the liquid refrigerant flow. Also, in low-temperature applications, the oil may freeze in the expansion valve area, and again cause operational problems.

As described in Chapter 2, a refrigerant blend is made up of two or more components. Because each component has different properties, the extent to which each component mixes with the lubricant will vary. For systems that have a small refrigerant charge compared to the oil volume, these differences can influence the physical properties of the blend and negatively affect system performance.

Types of Refrigeration Lubricants

There are several kinds of lubricants used in refrigeration systems. The most common are:

- Mineral oil (MO).
- Alkyl Benzene (AB).
- Alkyl Benzene/Mineral oil mix (AB/MO).
- Polyol Ester lubricants (POE).

Mineral Oils (MO)

Highly refined mineral oil (MO) has traditionally been used in systems with CFC and HCFC refrigerants.

Because of the different properties of HFCs, mineral oil cannot be used with these refrigerants—the miscibility and solubility for HFCs and mineral oil is different from that of CFCs and HCFCs. HFCs and mineral oils cannot be used together—oil return to the compressor and heat transfer in the evaporator and condenser are negatively affected.

Mineral oils cost about $12.50 per gallon.

Alkyl Benzene (AB)

Alkyl Benzene (AB) oils have been introduced in recent years (Zerol is one example). AB is one half synthetic oils and one half natural oils.

Specific AB brands are recommended for use as mixtures with mineral oil when using interim blends such as R-401A, R-401B, R-402A, and R-408A (MP-39, MP-66, HP80 and FX-10). (See "Alkyl Benzene/Mineral Oil Mix" below.)

AB oil costs approximately $16.50 per gallon.

Alkyl Benzene/Mineral Oil Mix (AB/MO)

Another common lubricant is a mixture of Alkyl Benzene and mineral oil (AB/MO).

If this mixture is to be used in a retrofit situation, virtually all of the existing mineral oil must be drained before the system is refilled with these products. This is done to make sure that the oil has a minimum of 50% Alkyl Benzene. (Specifically, the interim blends R-401A and R-402A require a minimum of 50% Alkyl Benzene oil in the system to make sure the oil circulates properly with the refrigerant and to assure proper oil return. A one-time oil change is sufficient.)

AB/MO is acceptable with traditional CFCs, is preferred with interim HCFCs, but not acceptable with HFCs.

Polyol Ester Lubricants (POE)

Neopentyl polyol esters, or simply Polyol Ester oils (POE), are typically used with HFC refrigerants. POEs have been used as jet engine lubricants for many years because they have a wide operating temperature range, are very stable, and lubricate well.

Polyol Ester oils can be mixed with mineral oils when used in systems with traditional refrigerants or interim blends. POEs are "backward compatible." That is, they are compatible with traditional refrigerants such as R-12, R-22, or R-502. Many POE lubricants have excellent miscibility with HFC refrigerants and conventional refrigeration oils.

POEs are derived from different substances, which means their physical properties and composition can vary widely depending on what exactly is used to make them. Also, POEs require additives to improve their performance. Additive packages contain substances that coat metal components to reduce system wear, and prevent corrosion. For use with HFCs, special additives also are necessary for stability (the different compressor manufacturers recommend somewhat different formulations).

Systems with HFC refrigerants cannot tolerate mineral oil as a lubricant, even in small quantities. This means when an existing system is converted to use HFCs, the mineral oil must be removed and replaced by a Polyol Ester oil for an acceptable miscibility between HFC refrigerants and the lubricant. The maximum amount of acceptable residual mineral oil is 5% or less. To reach this concentration, from three to five oil changes over a period of weeks are necessary. (See "Oil Changing Procedure" below.)

This indicates that for conversions from R-12 to R-134a POEs are necessary. Any residual mineral oil can be removed by flushing the system with POEs. Any small amount of R-12 that may remain will have a minimal effect on the performance of the lubricant and refrigerant.

POEs should not be used as lubricants in ammonia refrigeration systems.

Polyol Ester oils are very hygroscopic, which means they will absorb a lot of water if given the chance. This attribute makes special handling essential to make sure excessive moisture does not enter the refrigeration system. If too much moisture is allowed to enter the oil, there is a risk of acid formation, which could attack the components in the system.

POEs together with some new refrigerants will often loosen deposits that have accumulated from system wear in the compressor and electric motor, and from external contaminants admitted during service. This effect could possibly lead to problems for converted systems. Deposits are dissolved and collect in the expansion valve, and can block the refrigerant flow. Both moisture absorption and the blocking problem can be prevented by installing an appropriate filter-dryer in the liquid line.

Synthetic oils are expensive, many times the cost of all-natural mineral oils. POE oil costs $50 per gallon or more.

Lubricant Compatibility

The table below shows a summary of lubricants and their suitability with different types of refrigerants.

Lubricant Compatibility

Lubricant	CFC	HCFC	HFC
Mineral Oil	Good	OK (R-22 only)	No
Alkyl Benzene	Good	Good	No
Alkyl Benzene and Mineral Oil mix (minimum 50% Alkyl Benzene)	OK	Good	No
Polyol Ester	OK	OK	Good

Table 4-5 Lubricant Compatibility

Oil Changing Procedure

As we have seen, when converting from CFC refrigerants to other types, it is often necessary to change the lubricant when changing refrigerants.

For conversions to HFCs and other refrigerants that do not contain chlorine (such as R-404A, R-507, or R-134a), all existing mineral oil must be removed and replaced by a Polyol Ester oil (POE). This procedure may require several repeated oil changes (from three to five oil changes over a period of several weeks) to reduce the residual mineral oil concentration to 5% or less.

Interim refrigerants that contain R-22, such as R-401A and R-402A, will require a one-time oil change to an alkyl-benzene oil (AB). There are some exceptions—R-408A and R-409A will work with mineral oil, but is not necessarily recommended.

Figure 4-8 below shows the steps required to change a CFC system to a CFC-free system. An oil change is done using the old refrigerant in the system. After an oil sample test is acceptable, then the old refrigerant is recovered, and the new refrigerant added.

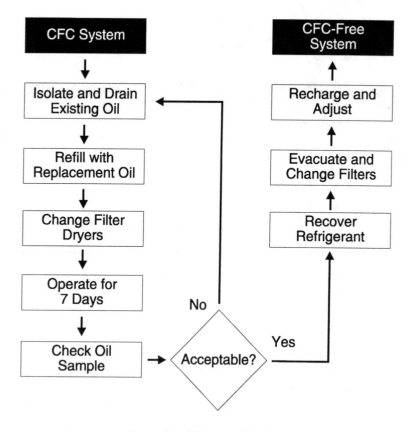

Figure 4-8 Oil Change Procedure

REFRIGERANT CHOICES FOR COMMERCIAL REFRIGERATION

This section presents information on some of the possible options for selecting a particular refrigerant and refrigeration equipment configuration. There are many options, since each situation will be different. For conversions, there may be a wide range of equipment, of varying ages and conditions. For new equipment choices, the application and other factors will dictate the choice.

This section provides information on:

- Short-term interim alternative refrigerants for conversions, specifically:
  - R-401A (MP-39).
  - R-402A (HP-80).
  - R-22.
  - R-408A.

- Long-term HFC alternative refrigerants (zero-chlorine options) for conversions, specifically:
  - R-134a.
  - R-404A.
  - R-507.
  - R-407A.
  - R-22 compared with R-404A and R-507.

- Retrofit options for existing systems including:
  - Existing R-12 medium-temperature systems.
  - Existing R-502 systems.

- Long-term options for new refrigeration equipment with engineered efficiency, including:
  - New medium-temperature systems.
  - New low-temperature systems.
  - Indirect refrigeration systems.

Interim Refrigerant Alternatives

The so-called interim refrigerant alternatives are options for the short-term since these refrigerants will, at some time in the future, be phased out. However, in certain retrofit situations, especially for refrigeration systems with some life left, and in situations where a store is remodeled often, it may make good economic sense to convert existing systems and use interim refrigerants.

A general rule of thumb is to use interim refrigerants for conversions and long-term refrigerants for new equipment. (One exception for new equipment may be to use R-22 with new equipment if it is anticipated the store will be remodeled and equipment will be replaced before R-22 is phased out.)

Using some interim refrigerants in a typical retrofit situation means lower first costs, but at the expense of higher energy costs. The energy efficiency of some of the interim blends is questionable. For some, very little on-site testing has been performed with sufficient instrumentation to be certain of the energy-efficiency impact. In most instances, before and after conditions are not compared; or, only run-hours are considered. Also, some of the interim blends have very high temperature glides, which could have unpredictable system effects and increase energy consumption substantially.

The principle interim refrigerants include R-22, a pure HCFC that contains chlorine, and refrigerant blends that all contain some R-22. With the exception of R-22, the predominant interim refrigerant choices below are strictly for refrigeration system uses only.

The interim choices predominantly available for refrigeration are:

- R-401A (MP-39), a near drop-in replacement for R-12.
- R-402A (HP-80), a near drop-in replacement for R-502.
- R-22, a possible replacement for both R-12 and R-502.
- R-408A, a drop-in replacement for R-502 with little glide.

R-401A

R-401A (MP-39), whose composition is R-22/152a/124 (53/13/34), is a near drop-in replacement for R-12. This refrigerant has a high glide temperature range (8° F) making performance reliant on the heat exchanger configurations in existing systems. In the testing at the SCE refrigeration lab (see "Commercial Refrigerant Research Project (A Case In Point)" later in this chapter) at 90° F saturated condensing temperature, R-401A was less efficient than R-12 by 2%.

R-401A is relatively easy to install in existing R-12 conventional systems. A single oil change to an AB oil and other minor system adjustments are required. The estimated conversion cost for a conventional store ranges from $12,000 to $15,000.

R-402A

R-402A (HP-80) is a near drop-in replacement for R-502. Its composition is R-125/290/22 (60/2/38). R-402A is noted to be 4% less efficient than R-502 in some literature. In the testing at the SCE refrigeration lab, at 90° F condensing, R-402A was 11% less efficient than R-502.

R-402A is relatively easy to install in existing R-502 conventional systems. A single oil change to an AB oil and other minor system adjustments are required. The estimated conversion cost for a conventional store is $12,000 to $15,000.

One issue regarding R-402A is its high operating pressure. It is not recommended for air-cooled systems in hot climates.

R-22

R-22 is an interim HCFC refrigerant that is typically known for its proven performance, and good efficiency. It has known characteristics, compressor performance, and oil compatibility. R-22 does experience high discharge temperatures which can affect equipment reliability. (The high discharge temperatures was one of the primary reasons for the development of R-502.)

R-22 has the advantage of being much less expensive than the blend alternatives. The initial low costs of R-22 and subsequent costs to replenish leaks can be attractive for certain systems. Potentially, the cost could increase on a supply-and-demand basis as production limits are imposed.

R-22 is not recommended for use in converting conventional systems.

- Converting R-12 systems to R-22 requires changes to the compressor, expansion valve, and piping. R-22 operates at higher pressures and has 50% more refrigerating effect than R-12. Existing compressors cannot typically be changed to R-22 from R-12. The lower mass flow with R-22 typically requires that suction line riser sizes must be reduced.

- R-502 systems converted to R-22 require suction desuperheating equipment. However, R-22 could be considered for certain R-502 parallel systems.

R-22 was essentially the only option available to the commercial refrigeration industry when the CFC phaseout began. Most users converted new low-temperature systems from R-502 to R-22. Many medium-temperature systems were already using R-22 or if not, were changed over.

Even as R-134a became available as a medium-temperature option, many applications such as supermarkets stayed with R-22 due to the larger physical compressor and piping sizes associated with R-134a. Also, R-22 has good efficiency characteristics on medium-temperature applications. For some medium-temperature applications, some of the new HFCs used to replace R-502 are not as efficient as R-22.

R-408A

R-408A, a blend of R-125/143a/22 (7/46/47), is an R-502 alternative being marketed recently by Elf Atochem.

R-408A has characteristics that justify further investigation, including lower operating pressures than R-402A and higher efficiency than R-402A and R-502. R-408A has a 1° F temperature glide. Although an oil change is not required for conversions, a change to MO/AB is recommended.

In a demonstration project conducted by Southern California Edison, lab results from tests run in October 1995 for R-408A are good. The results were interesting in that R-408A out-performed R-502 by 7 or 8% and demonstrated a 6% capacity increase.

Long-Term Refrigerant Alternatives

One alternative for commercial refrigeration systems is to use long-term refrigerant alternatives that do not contain chlorine (such as pure HFCs or blends that contain HFCs.) These alternatives may be used in either conventional or parallel applications, in new equipment (and in some cases, converted equipment).

For existing parallel systems with more efficient technology, it is recommended to keep existing equipment and apply new, long-term refrigerants when possible.

Refrigeration systems that implement long-term refrigerants have more expensive up-front costs, but will result in increased energy cost savings.

The long-term zero-chorine refrigerant options discussed below include:

- R-134a (pure HFC)
- R-404A
- R-507
- R-407A

R-134a

R-12 has one zero-chlorine replacement refrigerant: HFC-134a. This was the first HFC in large-scale production due to its selection by all major automotive manufacturers. Its cost has come down rapidly as production facilities have come on-line. R-134a is not a good alternative for new systems for practical reasons (compressor displacement and line sizes), but it could be considered for replacement of R-12 in existing medium-temperature systems instead of R-401A.

R-134a has similar pressures, thermodynamic properties and other characteristics as R-12. For medium-temperature commercial refrigeration applications, R-134a can be used as a retrofit replacement for R-12 with few problems. The primary requisite for a retrofit is the conversion from mineral oil to a POE oil, which requires multiple oil changes to reduce the residual mineral oil concentration to 5% or less.

The disadvantage of multiple oil changes could possibly be outweighed by the lower cost of the refrigerant over time and, arguably, the fact it is a "permanent" solution. However, when compared to R-12, an efficiency decline of 3% is noted by DuPont and the SCE lab tests show a decrease of 8% at 90° F condensing temperature.

R-134a was initially installed in a number of supermarkets as the first HFC installations in built-up commercial systems. These were medium-temperature applications, with R-22 being used for low temperature. When HFC blends became available to replace R-502 (and R-22) in low-temperature applications, it was immediately used for medium-temperature loads as well. The primary advantage of the higher-pressure HFC blends over R-134a is a reduction in system cost due to smaller compressors and line sizes.

The use of R-134a in new commercial refrigeration systems appears to be concentrated in small, medium-temperature systems and self-contained equipment.

Some elastomers used with R-12 are not compatible with R-134a. Also, compressors manufactured before 1975 have different motor winding insulation materials than later models, which may not be compatible with R-134a.

Some previous problems with R-134a conversions may be the result of high moisture levels or simply that poorly maintained systems were converted. The mixture of POE oil and HFCs tends to release existing system deposits which can contaminate compressors, valves, and other components.

R-404A

R-404A is an HFC blend composed of R-125/143a/134a (44/52/4) that replaces R-502 in both low- and medium-temperature applications. R-404A is one of the primary choices available for new commercial refrigeration systems and is approved and rated by major compressor manufacturers.

Initially introduced by DuPont, R-404A is now produced by DuPont, AlliedSignal and Elf Atochem. R-404A has a small temperature glide, but a relatively high global warming index which may impact its long-term viability.

Efficiency of R-404A is generally lower than R-502, particularly at higher condensing temperatures. DuPont shows a 10% lower efficiency (under standard conditions not representative of a supermarket). SCE lab testing showed an 8% lower efficiency at 90° F condensing. EPRI testing shows a 6% efficiency decline when compared with R-502, for an operating system that included mechanical subcooling.

Conversions to R-404A require a change to polyolester (POE) oil, including multiple oil changes to reduce the residual mineral oil concentration to 5% or less.

R-507

R-507 is a two-component HFC blend offered by AlliedSignal that replaces R-502 in low- and medium-temperature applications. Its composition is R-125/143a (45/55). R-507 is another of the primary choices available for new commercial refrigeration systems and is approved and rated by major compressor manufacturers.

SCE testing showed a 5% better efficiency with R-507 than R-404A, which other testing tends to support. However, many consider the two refrigerants to be approximately equal.

R-507 has no consequential temperature glide. R-507 requires a change to POE oil, including multiple oil changes to reduce the residual mineral oil concentration to 5% or less.

R-407A

R-407A is R-32/125/134a (20/40/40). A late entrant to the U.S. market, R-407A (KLEA-60) is of interest in two respects. First, it has lower global warming potential than either R-404A or R-507, which may be of importance in the future. Second, it is a high glide (10° F) refrigerant, which may offer improved efficiency if properly applied and controlled.

In certain water-cooled systems, the counterflow heat exchange properties of R-407 might have a definite advantage.

R-22 Compared with R-404A and R-507

As stated earlier, the HCFC R-22 in some situations might be considered as an "interim long-term" alternative in addition to R-404A or R-507. R-404A or R-507 are considered "long-term" alternatives. But which is the best choice?

R-22 could be a choice for new systems, especially for high-leakage systems. A change to R-22 can lower costs, especially for markets with a short remodel cycle (markets typically remodel every seven to 10 years). Also, HFCs may have problems not yet apparent.

After consideration of R-22 and R-134a above, the primary choices available for new commercial refrigeration systems are R-404A and R-507. Both were designed as low-temperature R-502 replacements but are being applied in low- as well as medium-temperature systems. In fact, nearly every supermarket chain that is using R-404A or R-507 for low-temperature systems is using the same refrigerant for the medium-temperature systems.

At the outset of the CFC phaseout, many companies rapidly moved to R-22 as an alternative to R-12 and R-502. R-22 has been used for many years, and while it has certain undesirable characteristics, the risks and challenges are well known.

Many chains continue to use R-22 in new construction with the following justifications:

- R-22 continues to be relatively inexpensive.

- R-22 system challenges are known, and HFCs still have a limited track record.

- Installation costs are reasonable (pipes are relatively small).

- Anticipates a future R-22 "drop in" replacement.

- R-22 phaseout is still years away.

While R-22 continues to be a good refrigerant that will probably be available for the next decade, transition to HFCs should now be considered for these reasons:

- R-22 cost could increase rapidly due to supply and demand pressures (production is limited).

- There are high discharge temperatures with single-stage R-22 systems (or use of liquid injection which affects reliability).

- R-22 systems with a two-stage configuration are more complex and have increased service costs.

- A zero-chlorine choice is desirable from an environmental perspective.

If use of R-22 is continued, refrigerant piping should be sized for the greater mass flows required by current HFC alternatives (such as R-404A or R-507)—which are close to the pipe sizes required by R-502. This way, whatever future alternative is used won't leave you unable to accommodate the pressure differences.

So, the best choice depends on many factors and the given situation.

Chapter 2 provides additional information on R-22 as it compares to R-404A and R-507 in air-conditioning applications.

Retrofit Options for Existing Systems

On the following pages are tables that provide some information on retrofit options for existing R-12 and R-502 systems.

These tables summarize the issues and considerations for changes from R-12 and R-502 to other refrigerants, comparing system characteristics, and giving some indication of how appropriately certain refrigerants might serve as replacement alternatives.

The emphasis is on the interim refrigerant alternatives. However, a retrofit might make use of long-term refrigerant if it makes sense for the given situation.

Existing R-12 Medium-Temperature Systems

Table 4-6 and Table 4-7 describe retrofit options for existing R-12 medium-temperature refrigeration systems.

Retrofit Options for Existing R-12 Medium-Temperature Systems		
Option Type	**R-12 Replacement**	**Issues and Considerations**
HCFC (Interim)	R-22	• Higher pressures, re-design required. • No oil change required. • New compressor required. • New expansion valves and (probably) distributors. • Usually requires reduction in suction riser size.
HCFC Blends (Interim)	R-401A (MP-39)	• Similar pressures, no compressor change. • 9° F temperature glide. • Single oil change to AB oil. • Requires change in dryer type. • No thermostatic expansion valve (TXV) change. • 15 - 20% capacity increase. • 2 - 3% efficiency decrease.
	R-402A (MP-66)	• Similar to 401A with higher pressures. • Suitable for some ice makers and lower temperature R-12 applications.
	R-409A (FX56)	• No oil change required. • + 2% capacity increase.

Table 4-6 Retrofit Options for Existing R-12 Medium-Temperature Systems

Retrofit Options for Existing R-12 Medium-Temperature Systems (continued)		
Option Type	**R-12 Replacement**	**Issues and Considerations**
HFC (Zero Chlorine Long-Term Alternative)	R-134a	• Similar pressures, no compressor change. • No temperature glide. • Multiple oil changes to POE oil. • Requires change in dryer type. • No TXV change required. • Check seal compatibility on valves, etc. • Systems with Carlyle compressors require an oil pump change. • Caution on pre-1975 compressors (check materials compatibility). • Typically no capacity problems; capacity remains the same. • 5 - 13% efficiency decrease.
Complete System Change	R-404A R-507 R-407A	• Requires new compressor system, expansion valves, and distributors. • Multiple oil changes to POE oil. • Reduce suction line risers. • May need to upgrade piping on air-cooled systems.

Table 4-7 Retrofit Options for Existing R-12 Medium-Temperature Systems (continued)

Existing R-502 Systems

Existing parallel systems are primarily R-502 systems. Generally, these experience the most frequent large refrigerant loss events, and for this reason, deserve special attention. The refrigerant choices for R-502 parallel systems include R-402A, R-404A, R-507 and R-407A.

On air-cooled systems, operating pressures definitely become a concern. R-402A in particular will exceed the rated pressures for the larger discharge lines on these systems. While the risk of tubing failure is probably low, the legal exposure should be avoided if choices are available that do not exceed the design ratings.

On parallel systems, floating head pressure more closely follows ambient conditions, requires a little more control, but saves energy. This option should definitely be implemented at the time of a retrofit. At lower head pressures, R-404A and R-507 appear to offer good efficiencies, essentially the same as R-502, and possibly better with medium-temperature applications.

An interesting alternative for existing R-502 medium-temperature systems is R-22. Pressures are lower, existing compressors will work, and the refrigerant cost is low. For systems with inherently high refrigerant loss rates, R-22 could be the best choice due to lower ongoing costs. There is some risk in this choice since R-22 has production caps that could trigger a supply and demand pinch long before its phaseout begins.

Retrofit Options for Existing R-502 Systems		
Option Type	R-502 Replacement	Issues and Considerations
HCFC (Interim)	R-22	• Higher pressures, re-design required. • No oil change required. • Compressor capacity usually adequate. • Requires new expansion valves and (possibly) distributors. • May require reduction in suction riser size.
HCFC Blends (Interim)	R-402A (HP-80)	• Somewhat higher pressures, no compressor change required. • Single oil change to alkyl-benzene oil. • Requires change in dryer type. • No TXV change required. • Copeland compressors need special relief valve. • Possible piping pressure rating problems on large air-cooled systems. • 10 - 12% capacity increase. • 6 - 16% efficiency decline.
	R-402B (HP-81)	• Similar to R-402A, but higher pressures. • Suitable for some ice makers
HCFC Blends (Interim)	R-408A	• Similar pressures, no compressor change. • 1° F temperature glide. • No oil change required, but MO/AB recommended. • No TXV change required. • 6% capacity increase. • 7 - 8% efficiency increase.

Table 4-8   Retrofit Options for Existing R-502 Systems

Retrofit Options for Existing R-502 Systems (continued)		
Option Type	**R-502 Replacement**	**Issues and Considerations**
HFC (Zero Chlorine)	R-404A R-507	• Similar pressures, no compressor change. • No or minimal temperature glide. • Multiple oil changes to POE oil. • Requires change in dryer type. • No TXV change required. • Check seal compatibility on valves, etc. • Systems with Carlyle compressors require an oil pump change. • Caution on pre-1975 compressors (check materials compatibility). • R-404A: minimal capacity difference; 2 - 12% efficiency decline. • R-507: slight capacity increase; +1% to -5% efficiency difference.
	R-407A R-407B	• Similar to R-404A and R-507 except R-407A and R-407B have a significant temperature glide (6 - 10° F) and may require additional expansion valve adjustments. • R-407A: 8 - 9% capacity decrease; 6 - 8% efficiency decrease. • R-407B: 1 - 2% capacity decrease; slight efficiency decrease.
Complete System Change	R-404A R-507 R-407A	• Requires new compressor system, system redesign. • Requires multiple oil changes to POE oil. • May need to change expansion valves and reduce suction line risers if adding subcooling. • May need to upgrade piping on air-cooled systems.

Table 4-9 Retrofit Options for Existing R-502 Systems (continued)

Options for New Refrigeration Equipment

Installing new refrigeration equipment opens the doors to many options and provides a "clean slate" with which to start. A conventional system and long-term refrigerant might be considered, or a new parallel system and a long-term refrigerant.

This section discusses some of the refrigerant and system options for new systems for new medium-temperature and new low-temperature systems.

New Medium-Temperature Systems

The process for selecting new equipment for medium-temperature systems is very similar to previous R-12 systems except that changes in compressor capacity, horsepower, and line sizing must be taken into consideration.

Options for New Medium-Temperature Systems		
Option Type	R-12 Replacement	Issues and Considerations
HCFC (Interim)	R-22	• Proven performance, good efficiency. • Size lines large enough for future HFC refrigerant. • Consult compressor manufacturer regarding oil selection for future HFC transition.
HFC (Zero Chlorine)	R-134a	• Requires POE oil. • Requires large displacement compressors and large line sizes similar to R-12. • Efficiency decline.
	R-404A R-507 R-407A	• Require POE oil. • May need to use heavy wall piping on air-cooled systems to meet Mechanical Code requirements.
	R-407C	• Similar to above with 6 - 10° F glide. • Potential efficiency gains with heat exchanger optimization.
Future Options	R-410A (AZ-20)	• Operates at very high pressures.
	R-717 (NH₃)	• Requires indirect systems, uses additional energy. • Many code issues not resolved.

Table 4-10 Options for New Medium-Temperature Systems

Although R-22 will be phased out in the distant future, it is listed here as an option for new systems since it may be a reasonable choice in some situations.

New Low-Temperature Systems

New low-temperature systems provide additional options for consideration. There are various selection criteria, including system efficiency, simplicity, initial cost, maintenance cost, and others.

Some options include:

- Single-stage systems with liquid injection. These can be used for either single units or parallel racks and existing compressor designs can be modified for liquid injection.

- Two-stage systems for industrial refrigeration system applications. These options can achieve maximum efficiency by incorporating intercooling and flash gas removal. There are higher initial costs, but two-stage systems are cost-effective on a life-cycle basis. Also, there will be higher maintenance costs due to additional components and control complexity.

- Internally compounded compressors used with either single or parallel racks.

Options for New Low-Temperature Systems		
Option Type	R-12 Replacement	Issues and Considerations
HCFC (Interim)	R-22	• Liquid injection or compound compressors or two-stage system recommended due to high discharge temperature. • Size lines large enough for future HFC refrigerant. • Consult compressor manufacturer regarding oil selection for future HFC transition.
HFC (Zero Chlorine)	R-404A R-507 R-407A	• Use POE oil. • May need to upgrade piping on air cooled systems to meet Mechanical Code requirements.
Future Options	R-410A (AZ-20)	• Operates at very high pressures.
	R-717 (NH$_3$)	• Requires indirect systems, using additional energy. • Difficulty with low temperature indirect fluids. • Many code issues not resolved.

Table 4-11   Options for New Low-Temperature Systems

Indirect Refrigeration Systems

One experimental, yet interesting long-term alternative system is an indirect refrigerant approach for use in supermarkets.

Instead of refrigerant piped directly to the fixture coils, refrigerant would be used to chill a brine, glycol, or other indirect fluid which would be pumped to the display cases through a secondary loop. Refrigerant charge for the entire store could be reduced to 300 pounds or less. Also, this type of system would bring improved efficiency when compared to new halocarbon refrigerants and would be halogen-free. Indirect cooling is an area of research already underway in Europe and is starting in the U.S. (to date it is only in the conceptual and experimental stages).

The primary challenge is to identify an appropriate indirect fluid that can be pumped economically. This technology will require new display cases and concerns include energy use required for longer pipes, pumping needs, etc. However in the not-too-distant future, indirect system retrofits could be a viable alternative.

Chillers that use ammonia (R-717) is one option considered for indirect systems. Ammonia chillers have made some progress in the air-conditioning chiller market. Ammonia is most easily used in water-cooled or evaporative-cooled configurations, but is also offered in an air-cooled design by one or more manufacturers.

Ammonia has a number of very attractive characteristics, including:

- Zero ozone depletion.
- Zero direct global warming (greenhouse) effect.
- Low cost.
- High theoretical efficiency.
- Low required mass flow per ton of refrigeration.

Ammonia in commercial refrigeration applications is limited by code restrictions because it is flammable and toxic. Proponents of ammonia are addressing these issues with research into indirect systems, low refrigerant charge, and special ventilation systems. Dealing with accidental release includes various methods to absorb the ammonia quickly before people are harmed. Possibilities could include diffusion systems (the ammonia is diverted to a water tank) or deluge systems (water is sprayed onto the ammonia and exposed area).

Traditionally, ammonia has been used in large plants. The lack of small, cost-effective compressors and other components suitable for small commercial systems limits somewhat the application of ammonia. Ammonia also has high discharge temperatures. Air-cooled systems therefore typically utilize screw compressors that mitigate the high temperatures using built-in oil cooling.

Copper cannot be used with ammonia, so all heat exchangers and system piping must use steel or aluminum tubing. Also, semi-hermetic compressors are not feasible, which means ammonia systems will need to use an open-drive compressor.

A system installed in a supermarket in Germany uses ammonia to chill a proprietary glycol fluid which cools both medium- and low-temperature display cases.

COMMERCIAL REFRIGERANT RESEARCH PROJECT (A CASE IN POINT)

In 1993, the Technology Planning and Development Department of Southern California Edison Company (SCE), Rosemead, CA, initiated a research project with the catchy title "Development and Demonstration of Energy-Efficient Commercial Refrigeration System Options using Non-CFC Refrigerants." The project included three phases—two phases were completed and reported on at the end of 1994.

The purpose of the research project was to develop environmentally compatible energy-conserving and demand-shifting commercial refrigeration systems using new non-chlorinated refrigerants, and to provide SCE's customers—principally supermarket and restaurant chains, and food-processing and cold storage facilities—with valuable research results addressing viable technical and implementation alternatives for regulation compliance and cost forecasting.

Project management was provided by SCE personnel, and the project consulting team included ASW Engineering Management Consultants, Tustin, California, and VaCom Technologies San Dimas, California.

The refrigerants that were tested included two CFCs, four HCFCs, and six HFCs. The following information is extracted from the December 1994 report on the project, and provides information on the lab configuration, testing procedures, results of lab and field tests, industry impact information, and more.

Purpose of the Research Project

The purpose of this research project was to develop and demonstrate optimized energy-conserving and demand-shifting commercial refrigeration system options utilizing new non-chlorinated refrigerants. Through this effort, the energy, environmental and economic impacts regarding the implementation of these systems were identified and communicated to SCE customers, manufacturers and other research organizations.

Without information on viable alternatives, most businesses will logically choose the first-available, lowest-cost solution that meets minimum CFC compliance, but which potentially could increase energy consumption and demand. Since commercial refrigeration represents a significant portion of SCE's total sales, an increase in energy use in this sector could require substantial investment in new generating capacity as well as cause owners to incur increased operating costs. For these and other reasons, it is in SCE's interest to participate in finding low energy-consuming refrigeration alternatives.

By developing and demonstrating more efficient retrofit alternatives during this "window of opportunity," SCE can assist its customers in making investment decisions that include optimal refrigeration systems options.

Project Results

This project included three phases.

* Phase 1—Applied Research: Evaluate and compare refrigerant alternatives on a fully-instrumented refrigeration test system.

* Phase 2—Analysis: Assess the viability of retrofit alternatives.

- Phase 3—Full-Scale Demonstration: Implement a full-scale pilot refrigeration system.

Project planning took place through 1992 and Phase 1 began operating the last quarter of 1993. Each test scenario operated up to two weeks for adequate data collection and the final analysis was completed in December 1994.

Phase 1 and 2 are complete.

Interested Edison customers have received a summary of the results of the research efforts to date. This information identified options that reduce energy consumption and operating costs, while eliminating CFC refrigerants. Specifics on different refrigerant and system choices, their environmental characteristics, economic factors, etc. were included.

Note: The lab is still operational. If anyone is interested in testing a refrigerant or other commercial refrigerant products, contact David Wylie at ASW Engineering Management Consultants at (714) 731-8193, 2512 Chambers Road, Suite 103, Tustin, CA 92680, or e-mail at asw@ix.netcom.com.

Phase 1—Applied Research

The first phase involved initial screening tests of refrigerants using supermarket display cases and associated refrigeration equipment. The test lab was constructed at Southern California Edison's Highgrove Research and Development Test Center near Riverside, California. The test system was used to evaluate and compare refrigerant alternatives for both medium- and low-temperature systems.

The data obtained from the test site permits efficiency comparisons for each refrigerant under "real" commercial system conditions. Actual food display fixtures operated in an environmentally controlled space, with a refrigerant compressor equipped with variable speed drive used to simulate the various volume displacements required by different refrigerants. Test runs compared "baseline" CFC refrigerants with interim HCFC blends and HFC (zero chlorine) azeotrope and zeotrope blends.

The test set-up included a heat rejection chiller which provided the capability to operate from 120° F to 40° F condensing temperature. Extensive instrumentation acquired data over a range of condensing temperatures as well as establishing the effects of variable speed and return gas temperature for normalization of test results.

The refrigerants that were tested are listed in the following table:

Tested Refrigerants		
CFCs	HCFCs	HFCs
R-12	R-22	R-134a
R-502	R-401A (SUVA MP39)	R-404A (SUVA HP62)
	R-402A (SUVA HP80)	R-507 (AZ-50)
	R-408A (FX-10)	R-407A (KLEA 60)
		R-407C (KLEA 66)
		R-410A (AZ-20)

Table 4-12   Tested Refrigerants

Later in this report are a description of the laboratory, test procedures, and specific results.

Phase 2—Analysis

Phase 2 of the project ran concurrently with Phase 1. This phase assessed the impact of retrofit alternatives for SCE customers who have conventional CFC systems including many supermarkets, restaurants, and cold-storage facilities. The choices considered were:

- Conversion of existing systems to interim blends, R-401A for R-12 and R-402A for R-502.

- New compressor systems, condenser, and conversion to HFC refrigerant.

- New compressor systems, HFC refrigerant, and low-temperature cascade condensing using off-peak thermal energy storage (TES).

Each of these alternatives was assessed based on:

- Capital cost estimations.
- Annual energy use modeling.
- Demand impact.
- Owning and operating cost estimations.

Systems that were tested included a conventional closed-loop system; parallel air- and evaporative-condensers; and a parallel system with TES and evaporative condensers.

Using field-installed instrumentation, SCE has collected electric-load data from many different supermarkets for several years. This data, a statistical summary of chain-wide equipment populations provided by a well-known supermarket chain, along with selected research results gathered in Phase 1, was used in computer modeling to evaluate and analyze system energy use for several replacement options for a single prototype supermarket.

The "prototype" store was medium sized at 32,000 sq. ft. and was assumed to use conventional (one-on-one) compressor systems with R-12 on medium-temperature and R-502 on low-temperature systems. (The "prototype" store is described in detail in a later section "Refrigerant Impact Analysis.")

A Baseline energy simulation was performed for the prototype store in two weather locations, Inland and Desert.

Using the refrigeration cooling load profiles from the Baseline design, the general CFC replacement choices (described above) were evaluated.

Each of these was considered in both weather locations. The alternatives involving new compressor systems were considered with both air-cooled and evaporative-cooled condensers.

Complete engineering designs were performed for each system and project costs estimated, including consideration of logistics to keep the store in operation while systems were changed.

Included later are tables of life-cycle analysis which show the relative economic value of each retrofit alternative and the overall financial impact to implement the refrigerant phaseout policy, including replacement methodology. Also included are energy use profiles with interim replacements.

Phase 3—Full-Scale Demonstration

In Phase 3, subject to interest and funding, a full-scale pilot refrigeration system may be implemented at a supermarket, using the most promising refrigerants from Phase 1, and new technology designed and developed in Phase 2.

Future research options could include additional component testing to identify if greater potential for high-glide refrigerants is achievable. This would include examination of evaporator modifications for optimum counterflow heat exchangers and development of electronic control valves.

A test site for Phase 3 (which would include redesign for HFC refrigerants and utilize thermal energy storage (TES) for subcooling) is being sought at this time and research funds are more than likely available. Anyone interested should contact David Wylie at ASW Engineering Management Consultants at (714) 731-8193, 2512 Chambers Road, Suite 103, Tustin, CA 92680.

Description of the Test Facility

The Commercial Refrigeration/CFC Demonstration Project test facility is owned by Southern California Edison Co. (SCE). The original project equipment was housed in a temporary building on a pad provided by SCE and was located at SCE's Highgrove Facility in Colton, California. (The project has since been moved to SCE's Customer Technology Center in Irwindale, California.)

Construction began in October, 1993 and was completed in January, 1994. Initial test runs began in February, 1994 and were completed in September, 1994. A subsequent test was run in October 1995 for R-408A. The interior of the temporary building was divided into two areas, separated by an insulated partition. At one end of the building, an area of approximately 14' W x 24' L was maintained as a "controlled environment," and housed the supermarket cooler and freezer display case fixtures. The remaining area provided space for the systems refrigeration equipment, data acquisition system scanners, and the man-machine interface computer.

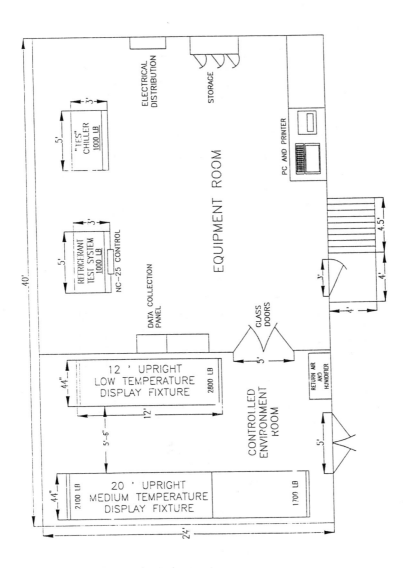

Figure 4-9 Test Facility

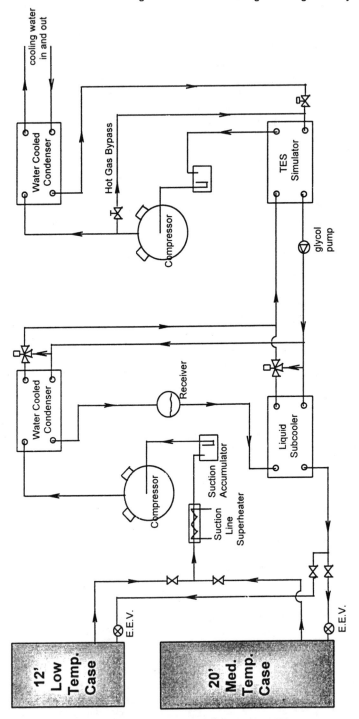

Figure 4-10 "Advanced" System Test Configuration

Review of Test System Components

The test system components described below include:

- Refrigerant Test Compressor Rack.
- TES Simulation Chiller Rack.
- Control System.
- Other Compressor System Materials.

Refrigerant Test Compressor Rack

Refrigerant Test Compressor Rack	
Major Components	
Compressor:	Carlyle 6.5 HP, Model 06DR820 semi-hermetic compressor
Condenser:	Alfa-Laval brazed plate-type condenser: Heat Rate 59,000 BTU/Hr Flow Rate: 870 Lb./Hr. R-12; 13 GPM glycol
Receiver Tank:	Standard Refrigeration, P/N UV-70, 10" diameter X 28" high vertical receiver tank
Subcooler:	Alfa-Laval plate-type subcooler: Heat Rate 12,400 BTU/Hr; Flow Rate: 870 Lb./Hr. R-12; 5 GPM glycol
Auxiliary Components	
Suction Accumulator	
Suction Filter	
Crankcase Heater	
Oil Fail Switch:	Electronic type
Suction De-superheating Expansion Valve	
Liquid Pump:	Smith Precision Products Model GC-1, equipped with a Smith Precision Products Model WW-120 Liquid Bypass Differential Regulator
Variable Speed Drive:	Danfoss YLT-3011 Variable Speed Drive for test compressor

Table 4-13   Refrigerant Test Compressor Rack

TES Simulation Chiller Rack	
Major Components	
Compressor:	Carlyle 6.5 HP, Model 06DR820 semi-hermetic compressor
Condenser:	Standard Refrigeration shell and tube condenser, Model # SST755-8 Pass
Chiller:	Alfa-Laval brazed plate-type chiller; Heat Rate: 66,900 BTU/Hr, Flow Rate: 977 Lb./Hr. R-22; 15 GPM glycol
Operating Conditions	
Glycol Temperature (In/out):	42°F / 32°F
Saturated Suction Temperature (SST):	20°F
Saturated Condensing Temperature (SCT):	105°F
Condenser Inlet Water Temperature:	85°F

Table 4-14   TES Simulation Chiller Rack

TES Simulation Chiller Rack

The test system's condensing and subcooling heat removal was provided through the use of a second custom compressor rack.

Control System

The control system for the project's refrigeration equipment included:

- One Danfoss NC-25 microprocessor-based control system, with analog input, analog output, and digital I/O boards.

- Software to communicate with the AKC case controllers and operate all test equipment. The program includes automatic sequencing through a series of condensing temperatures.

Other Compressor System Materials

Other compressor system materials included:

- 3/4" wall thickness AP Armaflex or Rubatex is used on all suction lines. All liquid lines, chilled water lines and heat reclaim lines are insulated with 1/2" AP Armaflex or Rubatex. Joints at elbows were mitered, and all joints were glued per manufacturer's recommendations.

- Three heating cables were provided for simulation of long return suction line: Heating cables are tightly attached to suction line with gear type clamps. Suction line and heaters are insulated with 2" fiberglass insulation with PVC jacket.

Display Fixtures

Two supermarket display fixture line-ups were used, one low temperature line-up 12' in length and one medium temperature line-up 20' in length (consisting of two separate fixtures, 12' and 8' in length).

Display Fixtures	
Low-Temperature Line-up	
Case	Tyler D6F124; four rows of shelving, 22" wide
Lights:	430 ma fluorescent, two rows in canopy, one row nose light
Defrost Type:	Electric
Case Controllers:	Danfoss "Adapt-Kool" electronic case controllers. Components: • One Danfoss AKC 16M Master Controller • One Danfoss AKC 16S Slave Controller • Two Danfoss AKV 12 Valves
Medium-Temperature Line-up	
Cases:	Tyler DDCM8 and DDCM12; four rows of shelving, 20" wide
Lights:	800 ma fluorescent, one row in the canopy
Defrost Type:	Ambient
Case Controllers:	Danfoss "Adapt-Kool" electronic case controllers. Components: • One Danfoss AKC 16M Master Controller • Two Danfoss AKC 16S Slave Controllers • Two Danfoss AKV 12 Valves

Table 4-15 Display Fixtures

Initial Refrigerant Charge

The test rack was evacuated to an absolute pressure not exceeding 1,500 microns. The vacuum was broken to 2 PSIG with dry nitrogen. The evacuation process was repeated, again breaking the vacuum with dry nitrogen. A drier of the required size in the liquid line was installed, and the racks were evacuated to an absolute pressure not exceeding 100 microns. The systems were verified to hold a 100 micron vacuum for two hours. R-22 refrigerant was then installed in the system.

HVAC Equipment for Controlled Environment Area

Air handlers, humidifiers, and other equipment were installed to improve test control and to minimize the impact of outside variables that might affect refrigerant variables.

HVAC Equipment for Controlled Environment Area		
Air Handler Motor CFM @ ESP:	One Magic Aire Model 36 BH air handler (without coil) 3/4 HP, two speed 1500 @ 0.75"	
Humidifier:	Two Research Products AprilAire humidifiers, Model 1140. Humidifiers installed at floor level next to return air grille of the air handler	
Supply Diffusers:	Metal-Aire Mod Flo series, with four modular cores and perforated face plate	
Heat Reclaim Coil:	Dimensions:	12" Fin Height x 36" Fin Length
	Rows:	6
	Fins per inch:	10
	Heat of Rejection:	89,950 BTU/Hr (available)

Table 4-16 HVAC Equipment for Controlled Environment Area

Data Acquisition System

The primary objective of the project was to obtain data and use that data to make calculations concerning the operating performance of various refrigerants. The range of the results (overall efficiency/performance of the refrigerants in kW/Ton, etc.) was anticipated to be fairly narrow. Due to the expected "tight" grouping of the test results, high confidence in the accuracy of the data used for the efficiency/performance calculations is necessary. If the errors in the data acquisition system are too great, meaningful distinctions concerning the refrigerants tested in the course of this project cannot be made.

With the objective of minimizing instrument error and maintaining a high level of repeatability and accuracy in the data, careful attention was paid to the system design, and steps were taken to: 1) use sensors of the highest accuracy available, 2) eliminate sensor drift errors by use of redundant sensors, 3) utilize calibration standard instruments of the highest accuracy, 4) eliminate interference from power conductors and high frequency signals by double-shielding sensor leads, and 5) calibrate the system at start-up and again mid-way through the series of tests using a 3-point calibration method.

The system uses a data scanner designed to provide measurements of great accuracy, with special emphasis on temperature. The data acquisition system for the project was designed to be completely separate from and independent of the supervisory control computer, to ensure data collection was not compromised by control sequences' priority over data acquisition.

Major Data Acquisition System Components

The main components of the data acquisition system include the data scanner and the man-machine interface.

Major Data Acquisition System Components	
Data Scanner	
Model:	Kaye Instruments Digi-4 Model #X1520S. Kaye's Digi-4 has a special emphasis on temperature measurement, with excellent thermocouple signal processing. The scanner was calibrated at factory, and is traceable to NIST standards.
Analog inputs:	• Fifty-four special grade T thermocouple inputs (+/- 0.03°C.) • Fourteen precision 100 Ω Platinum RTD inputs (+/- 0.01°C.) • Twenty analog inputs from pressure and other various transducers.
Outputs:	RS-232 Communication link, one report every 10 seconds. (A report includes instantaneous values of all data points.)
Man-Machine Interface	
Model:	IBM-compatible 80486-based PC, running National Instruments' LabView 3.0 for Windows. A communications link was maintained with the data-scanning equipment at all times. The PC also provided the system with data storage and remote (modem) data access to the projects' test data.
General PC Functions:	• Communicated continuously with the data scanner (via RS-232), acquiring one full data record (all scanned inputs) every 10 seconds. The data was then averaged over a 2-minute period, and the 2-minute averages written to storage (hard disk). • Maintained modem communication capability (for contact by a remote computer) and when requested, transmitted user specified files to the remote system. • Provided graphical representation of system performance data, both real-time and historical. Real-time graphic displays consisted of 1) a schematic representation of the system, with the flows, temperatures, pressures, and other process data indicated on the screen, and 2), trend graph windows. The data to be displayed in the trend graph windows was user-configurable, with up to 18 variables in three separate graph windows.

Table 4-17   Major Data Acquisition System Components

Data Acquisition System Calibration

The data acquisition system was calibrated during system commissioning, and at week 23 of the 34-week test sequence. The calibration was performed on all temperature and pressure points, using a 3-point calibration method.

Test Runs

Twelve refrigerants were tested, as shown in Table 4-12 "Tested Refrigerants." All runs with CFC and HCFC refrigerants were performed using the alkyl-benzene (AB) lubricating oil installed at start-up. Before installing and testing the HFC refrigerants, the AB oil was removed from the system by successive draining and flushing with the new ester oil. Laboratory testing of oil samples verified a concentration of AB oil of less than 1%.

- Refrigerant Change-Over(s): The refrigerants were removed and installed as follows:

  - The entire system was leak tested.

  - Refrigerants were removed from the test rack using a combination of liquid and vapor recovery methods.

  - Drier block was changed. The entire test system was evacuated to 200 microns or less. The test rack was charged with the new refrigerant, and manufacturers recommendations for blends were followed to ensure that proper mixture fractions were maintained.

- Test Run Sequence: A typical test run required 24 to 48 hours of continuous data collection. The test run sequence normally occurred on a 6-to-10 day cycle, shown in Table 4-18.

Test Run Sequence	
Day 1:	Start system, balance valves, and begin data collection.
Day 2:	Review first day of data. If good, continue run.
Day 3:	Review data run. If complete, proceed; if not, re-run test.
Day 4:	Perform speed and return gas temperature tests; continue review of test run data.
Day 5:	When all tests have been completed, remove and reinstall refrigerant.
Days 6 - 10:	Start-up with new refrigerant, adjust settings, verify operations required, etc.

Table 4-18   Test Run Sequence

The refrigerants tested, and the chronological order of testing, are shown in Table 4-19.

Tested Refrigerants and Sequence of Testing					
Week	**Run**	**Test Run Dates**	**Refrigerant**	**Oil**	**Fixture**
1 - 2	1	3/23/94 - 3/2594	R–22	Alkyl-Benzene	Low Temp
3 - 4	2	4/8/94 - 4/10/94	R–22	Alkyl-Benzene	Med. Temp
5 - 8	3	5/2/94 - 5/4/94	R–502	Alkyl-Benzene	Low Temp
9 - 10	4	5/13/94 - 5/16/94	R–12	Alkyl-Benzene	Med. Temp
11 - 12	5	5/23/94 - 5/25/94	R–401A (MP-39)	Alkyl-Benzene	Med. Temp
13 - 15	6	5/27/94 - 5/29/94	R–402A (HP-80)	Alkyl-Benzene	Low Temp
16		———————— Conversion to Polyol Ester Oil ————————			
17 - 18	7	6/17/94 - 6/19/24	R-134a	Polyol Ester	Med. Temp
19 - 20	8	6/24/94 - 6/26/94	R–404A (HP-62)	Polyol Ester	Med. Temp
21 - 22	9	7/8/94 - 7/10/94	R–404A (HP-62)	Polyol Ester	Low Temp
23 - 24		———————— Sensor Re-calibration ————————			
25 - 26	10	8/3/94 - 8/5/94	R–404A (HP-62)	Polyol Ester	Low Temp
27 - 28	11	8/16/94 - 8/18/94	R–507 (AZ-50)	Polyol Ester	Low Temp
29	12	8/26/94 - 8/29/94	R–407A (KLEA-60)	Polyol Ester	Low Temp
30 - 31	13	9/2/94 - 9/4/94	R–407C (KLEA-66)	Polyol Ester	Low Temp
32 - 33	14	9/9/94 - 9/11/94	R–410A (AZ-20)	Polyol Ester	Low Temp
	15	10/95	R-408A (FX-10)	Alkyl-Benzene	Low Temp

Table 4-19   Tested Refrigerants and Sequence of Testing

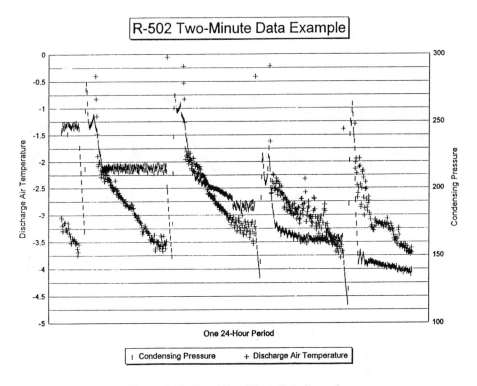

Figure 4-11   R-502 Two-Minute Data Example

Two-Minute Data Example

The graphic above shows sample data acquired on a two-minute basis indicating condenser pressure and discharge air temperature. The overall graph portrays a 24-hour period. The trends represent four, six-hour test runs followed by a defrost cycle. Discharge air temperature was held fairly close (+ 1° F) while condenser temperature was lowered 20° F for each six-hour cycle.

Test Result Tables

The results of the refrigerants tested are shown in the following tables. Using classical error analysis methods, the maximum error of the numbers in these tables is 1.5%.

Test Results 1

Medium-Temperature Efficiency Comparison at Selected Saturated Condensing Temperatures (SCT)
kW/Ton at 30° F Fixture Discharge Air Temperature

Refrigerant	70° F SCT		90° F SCT		110° F SCT	
	kW/ Ton	% vs. R-12	kW/ Ton	% vs. R-12	kW/ Ton	% vs. R-12
R-12	0.88		1.16		1.48	
R-22 (a)	0.79	-10.7%	1.11	-4.3%	1.50	1.7%
R-401A (MP39)	0.90	2.2%	1.18	2.0%	1.53	3.6%
R-134a	1.00	13.5%	1.25	8.4%	1.55	5.1%
R-404A (a)	0.82	-6.6%	1.21	4.5%	1.70	17.8%

(a) Note that R-404A and R-22 are higher pressure refrigerants than R-12, R-401A and R-134a.

Table 4-20 Test Results 1

Test Results 2					
Low-Temperature Efficiency Comparison at Selected Saturated Condensing Temperatures (SCT) kW/Ton at at -5° F fixture Discharge Air Temperature					
Refrigerant	70° F SCT kW/ Ton	% vs. R-502	90° F SCT kW/ Ton	% vs. R-502	110° F SCT kW/ Ton	% vs. R-502
R-502	1.44		1.90		2.43	
R-22	1.57	9.5%	2.12	12.0%	2.74	12.5%
R-402A (HP80)	1.53	6.4%	2.11	11.3%	2.81	15.5%
R-404A (HP62)	1.47	2.2%	2.04	7.6%	2.74	12.6%
R-507 (AZ-50)	1.43	-0.7%	1.94	2.3%	2.56	5.4%
R-407A (KLEA60)	1.55	7.8%	2.03	7.1%	2.59	6.4%
R-407C (KLEA66)	1.61	12.0%	2.08	9.7%	2.61	7.3%
R-410A (AZ-20)	1.35	-6.1%	1.80	-4.9%		

Low temperature R-22 results include the effect of liquid injection cooling at the compressor that may have had greater impact than in many actual systems.

For comparison purposes, an average of bubble point and dew point was used as the comparison saturated condensing temperature for refrigerants that exhibit a temperature glide.

Electronic expansion valves performed poorly with R-407A and R-407C due to temperature glide. Results would be somewhat better with proper valve control.

Table 4-21   Test Results 2

Within a year of completing the initial test runs, Elf Atochem, a refrigerant manufacturer, was interested in a comparison of R-408A with R-502. Arrangements were made and the lab was operated to perform this test in October, 1995.

The results are interesting in that R-408A out-performed R-502 by 7 or 8%. The other R-502 interim refrigerant tested (R-402A) had an energy increase of 6 to 15% over R-502. Prior to this test run, it was thought that most interim refrigerants for R-502 would have an increasing energy impact on the systems they were used in. This apparently is not the case.

Test Results 3					
Low-Temperature Efficiency Comparison at Selected Saturated Condensing Temperatures (SCT) kW/Ton at at -5° F fixture Discharge Air Temperature					
Refrigerant	**70° F SCT**		**90° F SCT**		**110° F SCT**	
	kW/ Ton	**% vs. R-502**	**kW/ Ton**	**% vs. R-502**	**kW/ Ton**	**% vs. R-502**
R-502	1.47	0.00%	1.95	0.00%	2.52	0.00%
R-408A (Forane FX-10)	1.35	-8.16%	1.80	-7.69%	2.33	-7.54%
For "glide" refrigerants, kW/ton results are adjusted to equate SCT to (dew point + bubble point)/2; October 25, 1995						

Table 4-22 Test Results 3

In the preceding tables, the efficiency difference between refrigerants can be seen to change with changes in condensing temperature. In operation, most systems operate at varying condensing temperatures through the year.

Actual systems may include differences—pressure drops, effect of subcooling, suction line heat gain, floating head pressure, etc.—which would change the results indicated above. It should be noted that while some efficiency differences above are significant, *system design* can have a greater impact on efficiency than the choice of refrigerant.

Refrigerant Impact Analysis

Using field-installed instrumentation, SCE collected electric-load data from many different supermarkets for several years. This data, a statistical summary of chain-wide equipment populations provided by a well-know supermarket chain, along with select research results gathered in Phase 1, was used in computer modeling to evaluate and analyze system energy use for several replacement options for a single prototype supermarket.

Introduction

For evaluation purposes, a representative "prototype" store was developed that included characteristics of the CFC-containing stores being studied. The Prototype Store is not identical to any particular store and is somewhat larger than the average of the total population.

The main reason for establishing a Prototype Store and associated refrigeration systems was to simplify the impact study due to the large number of stores (84) of a major 100-store supermarket chain within the SCE's territory that presently use CFC-12 and R-502 refrigerants. This was accomplished by creating a project database using all the available data that the supermarket chain has on file, then, the necessary information was extracted to define the Prototype Store's (baseline) refrigeration systems and other related parameters. The results of this effort are presented below.

Background

This section presents the database of the refrigeration system alternatives selected for their evaluation. The database consists of:

1. General parameters of the prototype store.

2. Prototype store general floor layout.

3. Prototype store's refrigeration equipment floor layout.

4. Identification of the current refrigeration systems (baseline) equipment.

5. Associated system diagrams.

6. The same information as listed in items 4 and 5 for each of the selected refrigeration system alternatives.

The information presented in this section was used to model the operation of the current refrigeration systems (baseline) and the recommended alternatives to identify the annual electric-energy consumption by each alternative and to establish retrofit implementation budgets.

Description of the Prototype Store

1. The store has a total area of 32,000 square feet with 23,000 square feet dedicated to the sales area.

2. The store has "one-on-one" conventional refrigeration systems; each system has its own compressor, liquid condenser, and associated refrigeration pipe.

3. The store has a refrigeration heat recovery system consisting of a heating hot-water coil in the main air handling unit serving the

store. This heating coil is piped to the water-loop of the fluid cooler serving the liquid condensers.

4. During space heating requirements, the condenser water loop temperature is raised to provide adequate heat transfer at the heating coil in the main air handler.

5. The identified refrigerated load groups and design load requirements are listed in the table below:

Prototype Store Refrigerated Load Groups and Design Load Requirements		
Load SST*	Cooling Load (MBH)**	General Remarks
- 35	56.7	This load consists of 120 ft. of coffin ice cream and juice cases.
- 25	235.0	This load consists of 120 ft. of multi-deck frozen food and one walk-in freezer.
10	47.8	This load consists of 96 ft. of self-service meat cases.
15	94.8	This load consists of 120 ft. of deli cases.
20	310.9	This load includes approximately 96 ft. of service meat cases, 32 ft. of beverage cases, 96 ft. of dairy cases and 120 ft. of produce cases.
* SST = Saturated Suction Temperature ** MBH = Thousands of Btu per hour		

Table 4-23 Prototype Store Refrigerated Load Groups and Design Load Requirements

6. The medium-temperature refrigeration systems employ R-12 and the low-temperature refrigeration systems employ R-502.

7. All of the display cases are open.

8. Each low-temperature display case is equipped with electric defrost. Coffin cases defrost once per day and multi-deck freezers defrost every six hours.

9. Since open display cases cool and dehumidify the air in the store, supermarkets have a quite complex air conditioning (AC) load relationship. For this reason, it was decided not to model the operation of the entire AC-system in the Prototype Store, but rather to develop only the heating load portion, allowing proper evaluation of heat reclaim system differences. For the heating load model, an average store space temperature of approximately 71° F was used.

10. The store's electric-energy consumption is paid under SCE's GS-TOU Rate Schedule.

11. The floor plan below shows the general layout of a typical store as defined for the Prototype Store. Similar stores were visited for the purpose of gathering information about walls, glass and roof areas, general display cases layout, operating hours and establishing "on-going" space heating requirements for heat recovery from the refrigeration systems.

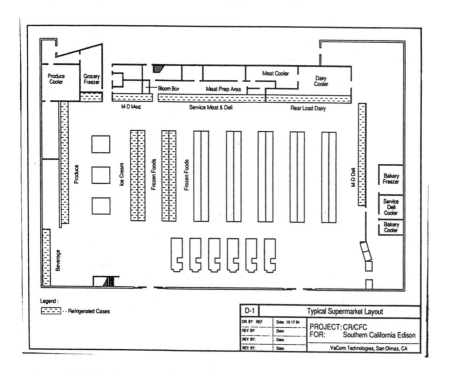

Figure 4-12 Typical Supermarket Layout

Alternative 1 (Baseline)—Current Refrigeration Systems

The Baseline refrigeration equipment for the prototype store consists of fifteen conventional compressor systems, using R-12 refrigerant on medium-temperature and R-502 on low-temperature systems. The systems are water-cooled and are connected to one fluid cooler utilizing evaporative cooling located in the machine room. The compressors are controlled by mechanical pressure switches. On most systems, the pressure switch control also effects fixture temperature control. On three systems with multiple loads, the fixture temperature is controlled with evaporator pressure regulator (EPR) valves and the compressor runs essentially all the time.

The water loop temperature is controlled by cycling a two-speed fan on the fluid cooler, with the fan either at high speed, low speed or off. It is also common to utilize a damper motor control on the fluid cooler that modulates airflow. In either instance, the water loop has two set points, one for normal operation and one for raised temperature during heat reclaim. Typically, the water temperature (entering the fluid cooler) is set at 90° F and 100° F, for the two operating modes.

The following drawing shows the typical arrangement of conventional compressor systems with individual condensers connected to a common fluid cooler.

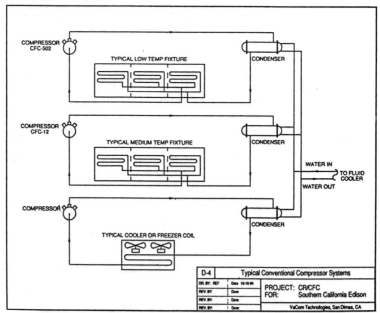

Figure 4-13 Typical Conventional Compressor Systems

Alternative 2—Interim Refrigerant Blends

This alternative utilizes the existing compressor systems with new refrigerants developed as "near drop-in" replacements. The compressors, expansion valves, line sizes, etc. are all suitable for the new refrigerants; 401A (MP-39) as a replacement for R-12 and 402A (HP-80) as a replacement for R-502. The compressor oil must be changed to Alkylbenzene synthetic oil (e.g. Zerol), however a one-time oil change is sufficient, since a portion of existing mineral oil (up to 50%) will not affect operation. Since the refrigerants operate at slightly different pressures and because 401A has nearly ten degrees of glide effect, controls must be adjusted to obtain proper operation.

Copeland compressors used with 402A in place of R-502 require the addition of a crankcase relief valve due to higher operating pressures.

The compressors, heat-rejection equipment and related controls remain the same as described above for the Baseline system.

Alternatives 3 and 4—New Parallel Refrigeration Systems

The parallel system design essentially includes a complete new machine room, with new compressor racks and controls, heat rejection and new air conditioning equipment. The equipment design was developed using the prototype machine room layout and considering the logistics necessary to keep the store in operation during the system change-out. The individual items to be installed include:

- Low-temperature parallel compressor rack.
- Medium-temperature parallel compressor rack.
- Evaporative condenser (or air cooled condensers).
- Branch system manifold.
- Defrost control panel.
- Vertical receiver/accessory packages (3).
- Temporary condensing units (2).
- Air conditioning compressor (located on XT rack).
- New heat reclaim coil for air handler.

All fixtures and walk-in loads are connected to the parallel systems, at the most appropriate design temperatures (saturated suction temperature or SST). The system is designed with five suction temperatures, -35° F, -25° F, +10° F, +20° F and +35° F. The +35° F level is used only for subcooling, not for any loads directly from the store. (Note that the

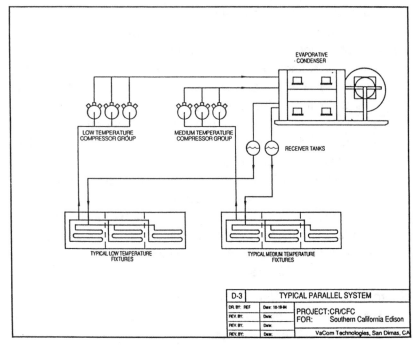

Figure 4-14 Typical Parallel System

supermarket chain does not use refrigerated preparation areas which might otherwise be connected to a +35° F system.) The above drawing shows multiple compressors connected in parallel.

The parallel compressor groups were selected using Copeland compressors on low- temperature systems, with at least three compressors in an uneven size combination for efficient cycling. The medium-temperature systems utilized Carlyle Compressor Co. compressors, with one machine equipped with unloaders (a method to run at low load which releases compression) to provide a large number of capacity steps with the fewest compressors.

Equipment selections were made for two weather locations and for both air-cooled (Alternative 3) and evaporatively cooled condensers (Alternative 4). The evaporative-cooled systems are the same for both locations. The air-cooled systems have slightly larger compressors and condenser selections for the desert location.

Refrigerant HFC-507 (AZ-50) was used for all parallel compressor selections. HFC-507 and HFC-404A (HP-62) are the only two refrigerants currently approved by Copeland and Carlyle Compressor Co. for use with their compressors. The choice between the two is arbitrary, since the performance is similar. Carlyle, for instance, offers only one set of compressor rating sheets, for use with either refrigerant.

HFC-404A and HFC-507 are refrigerants designed as R-502 replacements, whereas HFC-134a is a close match with R-12 and would ostensibly be the correct HFC choice to replace R-12 on medium-temperature systems. The higher pressure HFCs were selected over R-134a based on conversations with grocery chain engineers and industry experts indicating a limited potential for R-134a in new supermarket systems. The reasoning is that, due to its lower pressures, line sizes and compressor displacement are much larger than the higher pressure choices, with associated higher costs. The transition away from R-12 to R-22 or R-502 for medium-temperature loads resulted in considerable cost reductions and is doubtful the industry would retrace its steps. Also, chains evidenced a desire to use one refrigerant throughout the store rather than separate refrigerants for low and medium temperature.

Alternatives 5 and 6—New Parallel Systems with
Thermal Energy Storage

Alternatives 5 and 6 consist of air-cooled and evaporative-cooled versions of a parallel system design which utilizes ice-based Thermal Energy Storage (TES) as an integral part of the refrigeration system. To reduce electrical consumption and demand during high cost "on-peak" rate periods, the low temperature refrigeration compressors are maintained at a low discharge pressure by condensing into a "cascade condenser" connected to the TES ice bank. The glycol coming from the ice bank is at 36° F or less, allowing a condensing temperature of 42° F, with much lower compressor power than if the compressors were connected to a regular condenser.

The ice melted during the mid-day "discharge" period must be restored at night, during lower electric rate periods and, also, when ambient temperatures are lower. A glycol chiller is connected to a TES compressor that operates only during the TES "charging" cycle.

For the thermal energy storage alternative, several equipment and control configurations were considered, with the objectives including reliability, serviceability and energy efficiency. Since low-temperature products require uninterrupted refrigeration, it was considered necessary to develop a system that could continue to function without the TES system, should it fail to operate for any reason. This can be accomplished by selecting low-temperature compressors that can operate at high condensing temperatures if necessary or by designing a dual cascade approach, providing direct cooling by a medium-temperature system when TES is not being used. The latter configuration (aside from the TES portion) is not greatly different from the R-22 two-stage systems in common use in recent years.

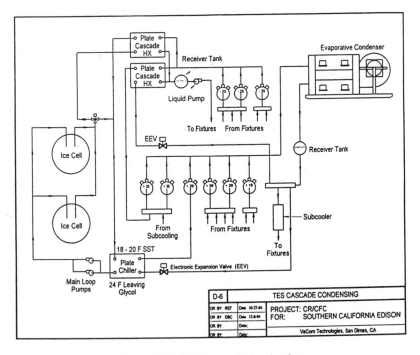

Figure 4-15   TES Cascade Condensing

The TES system includes the following major elements:

TES System Elements	
Cascade Low-Temp. System	The low-temperature parallel compressor rack is designed for a maximum condensing temperature of 42° F. Compressor ratings were developed by using Lab Testing results to extend the published manufacturer's ratings. With condensing at this temperature, additional liquid subcooling is not desirable. In order to deliver solid liquid to the low-temperature expansion valves, a liquid pump was included to add a relatively small pressure to the liquid (resulting in subcooling without heat removal).
Cascade Condensers	Two cascade condensers are included; one rejecting heat to the TES cells; the other transferring heat to the +35° F system. Both heat exchangers were designed at a close (6° F) approach temperature.
TES Chiller System	A single compressor is dedicated to the TES chiller for charging the ice cells. The compressor would be interconnected to the +20° F system, so that during initial start-up it would operate at a higher balance temperature but never go below +20° F saturated suction temperature. The TES chiller itself is designed as a close approach plate type heat exchanger (approximately 6° F approach—less at end of charge cycle).
Ice Cells	All plastic design ice storage cells were used to reduce weight and first cost. Two Calmac nominal 159 Ton-Hour cells were used.
Electronic Expansion Valves	The expansion valves on the TES chiller and +35° F cascade condenser are designed as electronic valves to achieve precise control at low approach temperatures. These are large tonnage valves using a fully analog technology—not the pulse width modulation method used with smaller valves.
Variable Speed Compressors	Variable speed compressor control was used in conjunction with fewer compressors to keep compressor racks at a manageable size and increase energy efficiency. This element could be included without the use of TES, of course, but is reasonably combined with TES since it is beyond the current technology commonly used in the Western U.S.
Liquid Suction Heat Exchangers	Specially selected high-efficiency liquid-suction heat exchangers were included on low-temperature and meat case systems for additional energy efficiency. This measure has been tested by EPRI with reasonable results.

Table 4-24 TES System Elements

Several different configurations were examined to explore the best approach for the TES design. The drawing previously presented shows the TES system configuration used for simulation and cost estimates. Other methods are possible and could be more effective for larger stores or under different utility rates. Since this system has not been built before, additional refinements would undoubtedly be found as part of a pilot TES installation.

Remodel Logistics

Supermarkets go to great lengths to remain open during major remodels. Certainly, any equipment changes have to be completed with no store closure or equipment down time. With equipment rooms frequently very tight, the logistical planning for a major equipment change-out can be complex. Installation costs are affected by how much room there is to work and how much work has to be done on a "rush" basis during night hours.

Using the typical prototype store, the remodel sequence was examined in detail to develop a plan that allowed room to work effectively and minimize costs. Since the compressors and heat rejection equipment are being removed, sequencing is a problem. The new compressors cannot run on the old fluid cooler, and the old compressors can not run on a new evaporative condenser. Plus, the new equipment must sit where the old equipment is now.

The solution was to employ two temporary single compressor racks; one for low-temperature loads and one for medium-temperature loads. These would be water-cooled units using city water to drain, or, if water supply was limited, connected to a temporary open cooling tower on grade behind the store. Suitable flexible connectors and other temporary facilities would be developed based on the intent to do more than one store with this approach. In conjunction, a remote manifold concept was developed to allow connection to the temporary units and subsequent easy connection to the permanent racks.

A side benefit of the temporary racks is to accomplish the oil conversion, from mineral oil to ester oil while running on the temporary systems rather than with either the old compressors or new racks. A means for doing this would be integral to the temporary racks, requiring very little labor compared to what would normally be required.

With remote manifolds, remote receiver tank assemblies and remote defrost panel, the system is more modular than traditional combined rack designs. In addition to reducing installation costs, this also allows greater flexibility when subsequent remodels require system changes.

Study Summary

Based on the research of the industry, which was performed during the development of this study and the identified financial results, ASW and VACom's conclusions are outlined below:

- The refrigeration system alternatives modeled with an evaporative condenser were, in every case, more energy-efficient than those with an air-cooled condenser.

- In today's market, there are "near drop-in" refrigerants available for existing refrigeration systems. Systems can be retrofitted with these refrigerants without having to invest in new refrigeration equipment. The identified "interim refrigerants" were R-401A (MP-39 DuPont and AlliedSignal), R-402A (HP-80 DuPont), and R-408A (Elf Atochem).

  The electric-energy models show that the use of these interim refrigerants would increase the compressor annual electric-energy consumption by as much as 11 percent over the current consumption. To implement this "interim approach" it was estimated that it would cost approximately $18,100 compared to $262,200 for a total retrofit for a store with similar characteristics as the "Prototype Store" defined in this study. These costs include an estimated $11,000 net CFC cost-avoidance credit, assuming the removed CFCs would be utilized in other facilities.

- The subsequent testing of interim R-408A revealed that an energy penalty is not always the case. Manufacturers of R-408A also claim that mineral oil is acceptable with this refrigerant. Although the impact analysis was completed prior to testing 408A, the conversion cost and operating cost can be somewhat lower than assumed with R-402A.

- In the application of replacement refrigerants, one should consider the increased design pressures as they relate to compressors, piping, and heat exchanger ratings. In addition, other code and safety issues may be applicable.

* Under system alternatives 6A and 6B, a thermal storage/cascade condensing system approach was evaluated. The low temperature cascade condensing would be affected by thermal energy storage as an integral part of the refrigeration system.

This system approach, while slightly more energy-efficient than 4A and 4B, will require approximately a 20% higher implementation cost than 4A and 4B. This system approach is not the most cost-effective system for the prototype store at this time; nevertheless, it should be studied further as incremental costs could be reduced. Performance improvements could be achieved as well from better refrigerant control and evaporator design.

Increasing costs due to utility demand charges will also have a significant impact on this option in the future. Additional performance and economic benefits could be realized due to the economy of scale inherent with larger stores. Currently, new stores being built in Southern California average 60,000 to 65,000 square feet, which is large when compared to the 30,000 square feet of the prototype store evaluated in this study. Depending on utility rebates, a TES system might achieve a 3 to 5-year payback.

The following life-cycle analysis shows the relative economic value of each retrofit alternative and the overall financial impact to implement the refrigerant phase-out policy, including replacement methodology.

System Alternatives and Annual Electric Cost					
Alternative	System	Condenser	Location	Refrigerant	Electric Cost/Year
1A	Conventional	Closed Loop	Inland	CFC 12/502	$ 77,559
1B	Conventional	Closed Loop	Desert	CFC 12/502	$ 76,785
2A	Conventional	Closed Loop	Inland	HCFC 401A/402A	$ 85,927
2B	Conventional	Closed Loop	Desert	HCFC 401A/402A	$ 85,078
3A	Parallel	Air	Inland	HFC 404A/507*	$ 54,742
3B	Parallel	Air	Desert	HFC 404A/507	$ 54,949
4A	Parallel	Evap	Inland	HFC 404A/507	$ 48,907
4B	Parallel	Evap	Desert	HFC 404A/507	$ 46,704
5A	Parallel-TES	Air	Inland	HFC 404A/507	$ 49,260
5B	Parallel-TES	Air	Desert	HFC 404A/507	$ 49,905
6A	Parallel-TES	Evap	Inland	HFC 404A/507	$ 45,864
6B	Parallel-TES	Evap	Desert	HFC 404A/507	$ 44,177

* HFC-507 was used for modeling, however R-404A and R-507 have similar efficiency.

Table 4-25 System Alternatives and Annual Electric Cost

Financial Analysis Summary					
Alternative	Implementation Budget	Annual Operating Cost	Present Value of 20-year Costs	Present Value vs. Alternate 1	Present Value vs. Alternate 2
1A	$ 0	$ 95,139	$ 1,315,302	—	—
1B	$ 0	$ 94,365	$ 1,304,639	—	—
2A	$ 18,100	$ 103,557	$ 1,424,394	$ 109,092	—
2B	$ 18,100	$ 102,708	$ 1,412,698	$ 108,059	—
3A	$ 282,400	$ 56,242	$ 1,202,988	($ 112,314)	($ 221,406)
3B	$ 291,000	$ 56,449	$ 1,214,440	($ 90,199)	($ 198,258)
4A	$ 262,200	$ 52,087	$ 1,123,595	($ 191,707)	($ 300,799)
4B	$ 262,200	$ 49,884	$ 1,093,243	($ 211,396)	($ 319,455)
5A	$ 323,800	$ 50,760	$ 1,180,525	($ 134,777)	($ 243,869)
5B	$ 329,100	$ 51,405	$ 1,194,711	($ 109,928)	($ 217,987)
6A	$ 317,800	$ 49,044	$ 1,148,934	($ 166,368)	($ 275,460)
6B	$ 317,800	$ 47,357	$ 1,125,692	($ 178,947)	($ 287,006)

Annual operating cost shown is for first year; yearly costs include assumptions for maintenance, refrigerant, and water. Costs are evaluated over a 20-year life assuming 8% capital cost and 4% energy escalation.

Table 4-26 Financial Analysis Summary

As shown in the tables, energy use with interim replacements actually increases somewhat. All parallel system options show reduced energy use with the evaporative condenser alternatives showing lower operating costs.

The cascade condensing/thermal energy storage alternatives show promise but are not particularly attractive for this size store (seven- to eight-year simple payback). This technology would show better financial performance on a larger store due to economies of scale and different utility rates associated with large stores.

EPRI RESEARCH (A SECOND CASE IN POINT)

The Electric Power Research Institute (EPRI) based in Palo Alto, California identified supermarket refrigeration as an area of interest and significant savings potential as early as the mid-1980s. With the assistance of Pacific Gas & Electric and other EPRI member utilities, Safeway Stores Inc., Hussmann Refrigeration and a large number of other industry participants, an extensive number of tests of alternative refrigerants and equipment combinations have been performed in an operating store located in Menlo Park, California.

Much of the work has been widely reported and recognized as a body of work unique in its practical application. Rather than being performed in a controlled laboratory, testing was subject to actual supermarket operations—including changes in store and outside conditions, shopping patterns, and typical maintenance problems.

The most recent body of work involved testing of R-404A, R-507, R-407A and R-407B, comparing performance with both R-502 and R-22. Certain combinations of heat exchangers and subcooling were included in the test, which may make the results quite specific to the test situation.

Overall, it was found that the differences between all of these refrigerants was quite small—often less than the calculated experimental error. One interesting conclusion was that R-407A showed a higher efficiency than R-404A or R-502, indicating that its glide characteristics may not have a substantial negative impact.

An important part of the EPRI research was testing of high-efficiency liquid-suction heat exchangers (LSHX). These heat exchangers were much like a standard LSHX except they were installed at the exit of the display case instead of inside the case, and were designed to have a much greater heat-transfer efficiency than the typical small LSHX. The heat exchangers improved efficiency with all refrigerants.

Other general conclusions reported from the EPRI testing include a recommendation that all refrigerant retrofit projects receive system adjustments (thermostatic expansion valve, etc.) as part of the work because significant performance improvements were seen in the test cases. Also, energy-saving enhancements were useful for the new refrigerants tested: floating head pressure and subcooling.

CHAPTER SUMMARY

Refrigerant options for commercial applications are greater than for air conditioning because commercial refrigeration systems require a wider range of operating temperatures and are more often custom-designed for specific installations. Also, unlike most air conditioning, the refrigerant lines can extend several hundred feet and are susceptible to leaks. Annual leakage for some systems can be greater than 15 to 20% of the charge which, for leaking CFCs, can be expensive. So, even though the scenario is the same: contain, retrofit, or build new systems, the commercial industry has considerably more equipment variables and refrigerant options.

Although refrigerant and compressor manufacturers have done some testing, side-by-side comparison of refrigerant performance in an actual commercial refrigeration system had not been accomplished with most of the interim or new HFC refrigerants.

EPRI embarked on a series of tests using an actual grocery store; however, the many variables associated with an operating store had to be accounted for, and the margin of error was sometimes greater than the difference detected from the refrigerants. However, the conclusion (that the differences were small) was in itself encouraging news for those forging ahead with system changes.

The testing conducted at the SCE lab attempted to eliminate as many of the system and usage variables as possible in order to see the actual characteristics of the refrigerant while doing the testing with real commercial hardware in a controlled environment. The results proved to be interesting. The impact of the performance of each refrigerant at various system conditions, helps clarify the economic choices available: contain, retrofit, or install new systems. And at many facilities, installing new systems presents the best economic option.

Chapter 5

Developing a Refrigerant Management Plan

INTRODUCTION

A plan to manage refrigerants outlines the best course of action to take concerning the transition from CFCs to environmentally acceptable refrigerants.

The planning effort requires a significant investment in your time and effort in order to:

- Do the research and analysis necessary to understand fully your alternatives.

- Compare what you are achieving now with what you will achieve with a retrofit or replacement using non-CFC refrigerant.

- Evaluate the trade-offs among your options.

- Select the course of action that is right for you.

Unfortunately, there is no set of "generic plans" you can choose from to achieve the best results. There are too many variables among different facilities, equipment, personnel, and other requirements. You must develop your own approach—one that is tailored to your specific situation and based on an individualized analysis and evaluation.

However, a comprehensive plan can yield great return on your investment of time and effort, and can help you avoid problems, minimize costs, and make decisions that make sense for today and for the future.

A sound plan will serve as the basis for making crucial decisions that can help you:

- Comply with regulations and restrictions relative to refrigerants, avoiding fines, penalties and possible lawsuits.

- Minimize capital costs.

- Reduce operating, maintenance, and other life-cycle costs.

- Ensure smooth operations now and throughout the refrigerant phase-out period.

If your plan is to do nothing and hope the issue resolves itself, you may:

- Face shortages of refrigerants (according to the EPA) as soon as 1996, possibly pay higher prices, and have trouble obtaining refrigerants.

- Experience long delays when you order retrofit parts or new equipment.

- Find yourself making hasty, uninformed, and potentially costly decisions regarding equipment.

In short, the longer you put off planning, the more you are increasing your risk and uncertainty—and the more expensive and disruptive it can be for you when you find yourself forced to take action.

There are five main steps to a sound approach for addressing the refrigerant phaseout issue:

- **Organize for Action**—put together the people and communication systems you will need throughout the planning and implementation process.

- **Analyze the Situation**—gather data about your current situation, including your equipment and constraints, to serve as a basis for making informed decisions.

- **Formulate an Action Plan**—evaluate your alternatives to determine the specific actions you should take, and when you should take them.

- **Execute the Plan**—proceed to implement improved maintenance and containment procedures, conduct conversions and retrofits, and purchase and install new equipment as outlined in the plan.

- **Monitor and Evaluate**—keep an eye on your progress and any changes (available technologies and regulations) that can have an impact on your plan to make sure you stay on track.

The graphic below illustrates these basic steps.

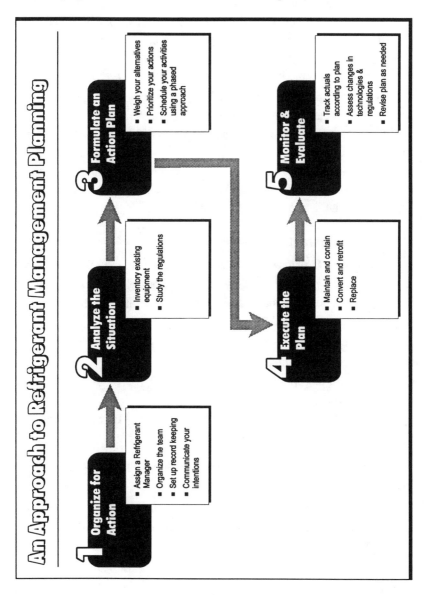

Figure 5-1 An Approach to Refrigerant Management Planning

Depending on your particular situation, the details for each step may be different. For example, your plan may include the details your organization wants to implement regarding employee training, educating management, financing equipment changes, procedures for tracking the plan and for reporting. At the end of this chapter is a review of how Unisys developed and executed their refrigerant management plan. Their experience may provide you with some valuable ideas.

ORGANIZE FOR ACTION

The first step for any project of this magnitude is to get organized. During the "Organize for Action" step, you set the stage for developing a sensible, comprehensive plan that will work within the context of your business environment.

Getting organized involves assigning a refrigerant manager, organizing the team that will be responsible for the plan, setting up recordkeeping, and communicating your intentions.

Assign a Refrigerant Manager

Assign a CFC manager—someone to spearhead the effort and oversee the whole operation—who will be responsible for establishing a CFC corporate policy position and a strategic operational plan.

This person will need to:

- Keep up with developments in the CFC arena.
- Analyze the existing systems.
- Understand the options available.
- Head up the planning process.
- Be in charge of writing a CFC management and transition plan.
- Make sure the plan is carried out.

The CFC manager will be responsible and accountable for refrigerant usage and conservation and should:

- Have clear authority and the necessary budget to effect change.
- Know about all of the facility's HVAC/R operations, industry standards, and relevant federal, state, and local regulations.

- Be able to coordinate and communicate successfully with other facility departments to ensure a coherent approach and smooth implementation of the plan.

- Have the confidence of upper level management and be able to gain their support and involvement.

Organize a Refrigerant Management Team

Depending upon the size of your organization and the number of facilities, the CFC manager is responsible for a refrigerant management team that may be organized to help facilitate planning and carrying out the desired actions.

It's probably a good idea to include a representative from an equipment manufacturing company as a team member.

Set Up Recordkeeping

If your company is not keeping good CFC records, it's never too late to start. Review the discussion in Chapter 1 under "Recordkeeping Requirements" that describes the kinds of records you should be keeping as required by Section 608 of the Clean Air Act Amendments.

In addition to meeting regulations, complete CFC records can help you identify areas where you can improve your CFC management and control procedures, and where you may need to tighten up your security for this very valuable and increasingly scarce commodity.

Communicate Your Intentions

To be successful, you need the understanding, support, and commitment of everyone who will be involved in making your refrigerant management plan a reality.

Employees and upper management alike need to have general knowledge of CFC phaseout issues in order to understand why taking action is crucial. They also need to know the general outlines of your plan, and what their role and responsibilities will be in making it happen.

You may want to work with key members of the refrigerant management team to draft a "mission statement" that outlines why the team has been assembled and what it is supposed to accomplish. This statement can serve as a starting place for communicating about the effort to others.

Staff members and outside contractors that deal with the HVAC/R system must have a detailed understanding of their specific responsibilities relative to refrigerant management. The approach taken by one CFC manager may work for you. He had internal staff and outside contractors sign an agreement promising they would conform to all the procedures set out in Section 608 of the Clean Air Act. To underscore the importance of this agreement, he posted relevant portions of Section 608 on each piece of equipment that contained refrigerant.

Upper-level management also must be aware of the key role they have. When they understand the external forces driving the changeover, they will see that it is a good business decision. This will make it much easier to get the funding and other resources you will need to implement your plan.

ANALYZE THE SITUATION

The next step of the refrigerant management planning process is to "Analyze the Situation." In this step you gather data about your equipment and the regulations and use this information to develop a sensible plan tailored to your specific needs.

Inventory Existing Equipment

To ensure that your plan encompasses all your refrigerant management needs you must collect data on all your chillers and other refrigeration machinery—including small appliances such as household refrigerators, window air conditioners, package units, vending machines, water coolers, and ice machines.

Typically equipment with the largest impact on your business—and the greatest refrigerant requirements—should take priority in the survey process. Therefore, it's often a good idea to begin the survey with equipment greater than 20 tons, then "work your way down" to the smaller equipment. This will help ensure that the most important bases are covered if you should find you do not have the time or resources to make the survey as detailed as you would like.

The check list on the following pages summarizes the specific information you should gather during this inventory.

Equipment Inventory Check List	
Data Category	**Information to Be Gathered**
General Background	• Buildings and locations. • Cooling capacity for each building or fixture.
Equipment Background	• Total number and type (scroll, reciprocating, centrifugal, etc.). • Manufacturer, model, model number, serial number, rated capacity and other nameplate data. • Date manufactured, year installed/age of the equipment. • Electrical power required. • Maintenance history of each compressor, including general service requirements and dates of last maintenance and recharge. • Next scheduled overhaul. • Condition of the equipment. Evaluating equipment condition takes more than an quick visual inspection. Review maintenance records and ask operations personnel about the equipment's overall performance and whether there have been recurring problems. • Anticipated life. In addition to the above evaluation of the equipment's condition, there are specific tests that can help you estimate its useful life. These include the eddy current test, oil analysis, and refrigerant analysis. (These tests are discussed under "Maintenance and Containment" later in this chapter.)
System Use and Output	System use and output data are key to assessing calculated and actual cooling loads and determining how they relate to the total system tonnage. If the equipment is connected to an energy management system, you can print a report that will give you data on its use and output over the period of a year. If the equipment is not connected to an energy management system, you can review the chiller logs to get this information.

Table 5-1 Equipment Inventory Check List

Equipment Inventory Check List (continued)	
Data Category	**Information to Be Gathered**
Refrigerant Use	Tracking the amount of refrigerant each piece of equipment uses will help you decide whether to maintain, retrofit, or replace a chiller. Keep logs, or inventory sheets, for all units, including information on: • Type of refrigerant. • Amount used, capacity in tons, and charge size. • Reclamation or recycling history of the refrigerant. • Service and disposal records. The example refrigerant log shown in Chapter 1 under "Recordkeeping Requirements" gives additional information on the type of information you should gather about refrigerant use.
Leakage Assessment	An assessment of the equipment's leakage rate can help you determine how much refrigerant you will need each year to keep a particular piece of equipment operating—and to support all the HVAC/R equipment in all of the buildings. Keep these figures up-to-date as equipment is retrofitted, converted, or replaced. You can get the leakage assessment data from records on recharging or "topping off," from leakage logs, or from leak detection/testing (discussed under "Maintenance and Containment" later in this chapter).
Refrigerant Stock/ Inventory	Because of requirements in the Clean Air Act, and because refrigerants are an important company asset, it is crucial that you have a documentation system that provides an "audit trail" for refrigerants. There are various methods you can use to organize and collect the data. For example, one method uses a system with a nine-digit number identifying each piece of equipment by location (building, section, floor), equipment type, and individual machine.

Table 5-2 Equipment Inventory Check List (continued)

Computer software also is available to help you manage your refrigerant inventory. For example, ASW Engineering Management Consultants in Tustin, CA can provide an economic planning tool in the form of a spreadsheet used with Lotus® 1-2-3 Spreadsheet for Windows (release 4). (You can contact ASW at (714) 731-8193.)

This tool can provide facility managers and planners with multiple-year forecasting and cost projections, and can help track refrigerant use and leak rates as well as refrigerant recovery and recycling history for multiple facilities. This template can be used to produce a variety of summary management reports and can generate data forms. In addition to helping with Section 608 compliance and supporting you in case of an EPA audit, this kind of software can help you uncover leaks that might otherwise go undetected.

DuPont, a major refrigerant manufacturer, also offers software programs designed to help facility managers and contractors. Their "Refrigerant Compliance Manager" software facilitates refrigerant management and inventory; and "Refrigerant Expert," provides information on the thermodynamic properties of DuPont products and can provide refrigeration cycle and piping design analysis.

Study the Regulations

To develop a plan that addresses your current and future refrigerant management needs, it's important that you understand all the regulations, standards, and laws governing the use of refrigerant—your local regulations and codes as well as those by the EPA, ASHRAE, ARI and others.

You must understand both the near-term and long-term outlook for the regulations—what's currently "on the books," what's pending or proposed, and what the regulatory trends are.

Also make sure you fully understand the costs and risks associated with non-compliance.

This book provides a good beginning in understanding the key regulatory issues, but this arena is marked by frequent change. It will take some effort on your part to make sure you keep up-to-date with the regulatory issues impacting your operation.

Consider Refrigerant Costs and Availability

The cost and availability of refrigerants, in particular CFC refrigerants, may be an important aspect of your overall analysis. CFC costs have risen dramatically over the past few years, due to new taxes and supply and demand forces that are driven by the scheduled production phaseout (no new R-12 or R-502 can be produced or imported). R-12 cost about $1.50 per pound ten years ago. R-12 and R-502 now cost around $10 per pound.

So, if your plan includes keeping equipment that uses CFC refrigerants, and if this equipment has a high leak rate, your maintenance costs could rise dramatically.

Also, if you decide to maintain existing CFC systems, you must be able to obtain supplies of CFC refrigerants after the phaseout. Although there are currently no widespread shortages of R-12 and R-502, there is concern regarding possible R-12 shortages because of the need for R-12 in automotive service and for centrifugal chillers used for air conditioning. The sheer number of automobiles with R-12 is a problem, and centrifugal chillers cannot be retrofitted at the pace necessary to avoid a shortage. These two uses will continue to tap R-12 reserves (at increasing cost) as long as they are available.

The actual amount of stockpiled refrigerant is not known, but there are no indications that large quantities of new refrigerant have been stored. Recovery efforts have provided some supermarket chains with a significant stock of used refrigerant.

R-502, used mainly by commercial refrigeration and over-the-road refrigerated trailers, may present a larger supply problem than R-12. R-502 has received less attention than R-12 and there is probably a much larger installed quantity of R-502.

There currently are ample supplies of HFC refrigerants to fill the need for R-502 alternatives.

R-22 (which is the only HCFC of concern to the supermarket industry) will be available under current regulations in at least service quantities until 2020. However, production is capped in 1996 at 1989 levels and production rates start to decline beginning in 2004.

The following table shows recent wholesaler costs for various refrigerants.

Refrigerant Costs	
Refrigerant	**Price Range**
R-12	$8.00 to $11.14 (includes $5.35 tax)
R-134a	$3.78 to $4.59
R-123	$4.00 to $7.00
R-22	$1.45 to $1.99
R-407C	Approximately $9.00
R-502	$8.20 to $10.05 (includes $1.64 tax)
R-401A (MP-39)	$4.33 to $4.72
R-402A (HP-80)	$6.76 to $7.21
R-404A	$6.69 to $8.49
R-507 (AZ-50)	$7.25 to $7.52
R-407A	Approximately $6.75
R-410A	Approximately $14.00

Table 5-3 Refrigerant Costs   *September, 1995*

As part of the analysis for supermarket and other commercial refrigeration systems, the key issues of refrigerant availability and cost should be considered, especially if there are many different systems.

Table 5-4 shows some average figures for a typical medium-size supermarket chain (100 stores). It includes a breakdown by type of compressor system (conventional and parallel), refrigerant and quantity, and annual refrigerant usage and leak rates. (Store chains should develop or use existing spreadsheet analysis tools so this kind of information is available.)

From the annual leak rate (%) given in the table, you can calculate the approximate expense caused by leaking systems. In this example, which shows typical values, the 500 conventional and 20 parallel R-12 systems have an average annual leak rate of 7%. This means the average medium-size supermarket spends around $60,000 (6,000 lbs. times $10) per year simply "topping off" existing R-12 systems. Also, the chain will need to have available a supply of at least 6,000 lbs. of R-12 for each year these systems are in use (provided the leak rate stays the same.) The kind of information included in the table below is especially critical for large chains with a number of stores.

Also, as part of your plan, you need to keep track of any refrigerant inventory that's available to feed existing stores over the near term. For example, if you phaseout R-12 systems and replace them with R-502 systems, you can track the recycled R-12 available for use in remaining systems.

Typical Medium-Sized Store Refrigerant Usage					
Refrigerant	Compressor Qty. Conv.	Parallel	Refrigerant Qty. lbs.	Annual Usage lbs.	Annual Leak Rate, %
R-12	500	20	85,000	6,000	7 %
R-502	600	100	100,000	15,000	15 %
R-22	100	100	30,000	9,000	30 %
TOTAL	1,200	220	215,000	30,000	14 %

Table 5-4　Typical Medium-Sized Store Refrigerant Usage

Consider Assistance from Outside Consultants

If you feel you may be getting in over your head in analyzing your situation, engineering consultants are experts in this arena and can help you in many ways. Companies like ASW Engineering Consultants can assist with implementing a CFC phaseout plan (you may call ASW at (714) 731-8193). They provide services such as:

- Energy and economic analysis.
- Project management.
- Retrofit design and specification.
- System start-up, specification verification, and fine-tuning.
- Data collection and analysis.

FORMULATE AN ACTION PLAN

When beginning to outline your refrigerant management action plan, it is a good idea to clarify your goals for the plan: exactly what is the plan to accomplish and how broad is its scope. Step three, "Formulate an Action Plan," addresses these issues.

At a minimum, the goals of the plan should include provisions for:

- Compliance with applicable laws and regulations.
- Continued refrigerant supplies and equipment service.
- Reduced refrigerant emissions.
- Increased recycling of refrigerants.

The plan should outline the approach you intend to take with each major piece of equipment over the next 30 years. It's generally a good idea to break the plan into phases reflecting immediate, near-term, mid-term, and long-term activities.

The immediate and near-term portions of the plan should provide details on the phase-in schedule, work items, and prospective dates for any replacements or retrofits you plan.

As you set your priorities and decide how to handle each issue, you need to weigh the pros and cons of the three basic alternative approaches you have available to you.

Three Alternative Actions

As introduced in Chapters 3 and 4, there are three basic approaches to addressing refrigerant issues:

- Maintain the existing equipment containing and preserving the refrigerant as best you can.
- Retrofit or convert the equipment so it can use alternative refrigerants.
- Replace the equipment with new models that are designed for alternative refrigerants.

Most plans use a combination of all three approaches, and the action taken is influenced by many factors including the age, condition, cost, and criticality of the equipment. The decision also reflects the equipment choices available, operating costs, any utility rebates that might be available, energy efficiency, environmental impact, and any required change in capacity if there are plans to remodel the facility.

Maintain and Contain

Maintaining existing equipment and containing and preserving the refrigerant as much as possible may be the best option for equipment that is less than 10 years old and in good working order.

It's perfectly legal to use CFCs in air conditioning and refrigeration equipment for as long as they exist, and there will probably be an ample—but expensive—supply of recycled refrigerant for years to come. (Carrier offers a program that guarantees operation of existing R-11 chillers until the year 2000, not only for their own chillers, but also for those from other manufacturers.)

Although this may be the least expensive approach, you need to consider that you may need additional (add-on) equipment to minimize leaks and to conserve existing refrigerants. If this is the case, you need to explore your options relative to the costs of these add-ons and their effectiveness over time.

Purified used refrigerant may be available from a certified reclaimer—subject to supply and demand. Used refrigerant can also be reused at any facility owned by the same owner as the facility from which it was removed. CFC refrigerants can be removed and reused in other systems belonging to the same owner. (See "Recovery, Recycle, & Reclaim" later in this chapter.)

If a system can be maintained leak free, the refrigerant charge can remain indefinitely. So, an option to consider with any system is how long the system can be maintained by simply reducing the leak rate.

The characteristics of a system where the "Maintain and Contain" option may be the best choice include:

- A system with limited life expectancy due to business expectations such as planned remodel, replacement, etc.

- Systems for which a change to alternatives would be expensive or might have an impact on reliability (such as high condensing temperature R-502 systems).

Containment efforts specific to AC and refrigeration systems are discussed in Chapters 3 and 4, respectively. A discussion on leak detection is provided under "Implementing a Leak-Detection System" later in this chapter.

Retrofit or Convert

As discussed in Chapters 3 and 4, if you have relatively efficient equipment that is in good condition and has a long lifetime ahead, retrofitting or converting the equipment to use newer refrigerants is a viable alternative. (However, as one source said, "It makes no sense to convert a dog into a converted dog.")

The options for retrofitting or converting range from relatively simple techniques—such as just changing the refrigerant and oil—to more complex and expensive approaches such as replacing the motor and compressor driveline assembly and installing new microprocessor controls.

Typically, the simpler, cheaper approaches take their toll in decreased system efficiency and lower cooling capacity—unless you take specific steps to upgrade equipment performance at the same time as the conversion.

Deciding whether to retrofit, and which level of conversion to use is a balancing act between the amount you want to spend and the operating performance you'd like to have. (This is very similar to measuring trade-offs when selecting a new, replacement system.)

Some of the specific issues you must consider when making a retrofit or conversion decision include:

- Maintenance history and condition of the equipment, especially piping.
- Energy efficiency of the equipment. A computer analysis can present options for energy and capacity trade-offs.
- Recovering refrigerant from converted equipment for use in other systems.

Replace

Replacement is often a good alternative for old or inefficient equipment, or for equipment that is especially crucial to the business.

While new equipment tends to have lower operating and maintenance costs—and usually has the advantage of accommodating future needs—it often requires significant initial investment. (However, these costs often can be paid back in a relatively short period, and finance and leasing options may let the system "pay for itself.")

As you think about replacing old equipment, remember that you will be able to reclaim or recycle the CFC refrigerant and use it to keep other machines operating.

Things to Consider as You Weigh the Alternatives

In developing the plan, you need to weigh your alternatives in terms of:

- Immediate, near-term, and long-term capital requirements.
- Operations, maintenance, and energy costs.
- Current and future refrigerant costs.
- "Before and after" conversion costs—including the costs of repair and retrofitting.
- Enhanced system equipment (add-on) costs.
- Possible exposure to risks.

(For example, what is the potential impact of a system failure with this specific equipment? What will be your alternatives if you are unable to get the necessary refrigerant? What penalties will you face if you are unable to make this equipment comply with regulations when necessary?)

Other issues that you should consider when developing your plan and making the maintain and contain, retrofit or convert, or replace decision include:

- The ODP and GWP of the current and replacement refrigerants.
- Refrigerant phaseout schedules and the various regulations governing your operation.
- The anticipated availability of refrigerants and equipment, and what alternatives will be available.
- The age and remaining life of the existing equipment and its maintenance and repair history.
- The amount of time the equipment is on-line.
- Trade-offs between capacity and energy use and efficiency.
- Any planned building modernization activities or planned equipment room modifications.
- Construction costs for new equipment.
- Energy efficiency rebates or other incentives from utility companies.

Concurrent Energy Saving Opportunities

As part of the analysis, keep in mind that there are several ways to mitigate the cost of converting a system to use another refrigerant, or the cost of installing new equipment. You can concurrently upgrade either chiller or refrigeration systems with features that will provide, in the long run, substantial energy savings—which can reduce the cost of conversions or new equipment. In addition, there are steps you can take at any time, whether you are changing refrigerant or not, that can reduce your overall energy costs.

For either refrigeration or air-conditioning systems, at the same time as you are changing refrigerant, take advantage of the opportunity to implement energy saving features such as:

- Variable-speed motor control.
- Floating head pressure.
- Microprocessor compressor control.

Other options that you can implement at any time, but that especially make sense if you are changing refrigerant or rebuilding either a refrigeration or air-conditioning system are:

- Install a multi-stage or multiplex cooling system with remote condensers.
- Convert air-cooled condensers to evaporatively cooled condensers.

There are other low-cost opportunities for both refrigeration and air conditioning not directly tied to a refrigerant change but that can be addressed at the same time. These are general procedures that may have benefits in terms of obtaining management interest or financing. These opportunities include:

- Maintain and fine-tune HVAC and refrigeration equipment, clean condensers and evaporator coils.
- Adjust expansion valve superheats.
- Implement lighting control.
- Install or retrofit facility lighting with more efficient lamps and ballasts, and consider electronic ballasts.
- Install variable-speed air handler and HVAC control.

Also, there are a few low-cost energy saving techniques and equipment additions specific to refrigeration systems and supermarkets that you could do at any time and are not dependent upon changing refrigerant:

- Adjust display case and walk-in boxes to proper temperatures.

- Follow manufacturers' operating and stocking instructions.

- Switch to demand defrost instead of automatic.

- Install humidity controls.

- Install anti-condensate heater control.

- Replace open cold cabinets with closed cabinets (a higher-cost alternative).

Setting Your Priorities

When selecting which equipment to convert or replace, you may want to look first at those that are:

- Relatively inexpensive to convert or replace.

  Open-drive compressors often can be converted by changing the seals and gaskets and with minor control modifications in addition to replacing the refrigerant. This type of conversion usually costs about half of what it costs to convert units with hermetic motors.

- Used in critical applications.

  Systems that serve functions that are critical to the business—such as process cooling or data center air conditioning—often make the "high priority" list for conversion or replacement.

- Already in need of major repairs.

  If a piece of equipment needs major repair, the cost of retrofitting/ converting or replacing is offset by the amount you would otherwise need to spend in order to fix it. In addition, the CFC refrigerant you can extract from the chiller can be considered an asset that can help defray the costs of converting or replacing the equipment.

- Able to take advantage of utility rebate programs.

  Often utility companies have rebate programs that you might be able to take advantage of to offset the costs of a major conversion or replacement.

EXECUTE THE PLAN

After you have organized for action, analyzed the situation and formulated an action plan, you are ready to "Execute the Plan." Now you begin to actually do the activities you set the framework for in the previous steps.

As you proceed with your plan, you may find you need to review the original details of the plan and make adjustments based on factors that you had not anticipated. In general, you should document these changes in your plan so you always have a comprehensive reference you can use to keep all team members informed, and to put your future activities in perspective.

As discussed earlier, your plan will basically encompass three kinds of activities: maintenance and containment, retrofits and conversions, and replacements.

Maintenance and Containment

If you have an assured source of affordable CFC refrigerant, you may choose to postpone conversion or replacement for many years. If you elect to maintain existing CFC-based equipment, you need to consider how you will accomplish the following:

- Improve operations and maintenance practices to make sure you are doing all you reasonably can.
- "Tighten up" all systems to minimize refrigerant losses through leaks.
- Control purges and pump-outs.
- Avoid catastrophic failures.
- Recover, recycle, and reclaim refrigerant.
- Stockpile or "bank" refrigerant to ensure that you will have enough to meet your needs.

Improve Operations and Maintenance Practices

Your general operational and maintenance policies and procedures can have a significant impact on how well you are able to keep your older equipment running efficiently with minimal refrigerant loss.

Some specific steps to help ensure effective operation and maintenance include:

- Clearly define containment practices—preferably through a written containment policy:
  - Register containment devices.
  - Establish training and specific service practices relative to containment.
  - Evaluate equipment for additional containment.

- Establish operations and maintenance schedules that specify who, what, when, where, and how equipment should be checked, tuned, and repaired.

- Make sure you are getting accurate and complete information about the equipment:
  - Calibrate gauges and thermometers regularly.
  - As a way to detect leaks, keep a log of refrigerant levels in the receiver. (Many, but not all air-conditioning and refrigeration systems have a component called a receiver that holds refrigerant and allows for expansion and contraction of the fluids. This is a good location to monitor the refrigerant level.)
  - Implement a leak-detection system (more on this topic follows).
  - Conduct oil and moisture analyses.

    The condition of AC/R system oil and the quality of the refrigerant can have a significant impact on system performance.
  - Evaluate your water treatment effectiveness.

    Effective water treatment will protect tubes, tube sheets, and water boxes against scale, fouling, and wear, and help ensure your system is working as it should.

    Make sure chemical feed pumps and sensing and control devices are functioning properly. Also consider collecting and analyzing water and deposit samples using corrosion-monitoring coupons.
  - Conduct other appropriate equipment tests, including:
    - Hermetic electric motor insulation testing to verify that the heat generated by the motor has a minimum negative impact on the system.

- Eddy current testing. This is an electronic, non-destructive method of inspecting heat exchanger tubes for defects.

- Vibration analysis to determine the vibration levels and sources, noise, and balance conditions for all AC and process refrigeration equipment. This information can help you avoid major problems that misalignment, bearing wear, or an out-of-balance condition can create.

- Infrared testing also can help identify potential problems by using heat-sensing scanners to find "hot spots" on electrical components, including wiring.

Implementing a Leak-Detection System

Halocarbon refrigerants are colorless and odorless and generally exist as vapors at room temperature and pressure. They are heavier than air and settle to low areas of a room. A system to monitor and detect leaks should be a part of your regular maintenance and repair program, and must include a log of all leakage data.

Detecting refrigerant leaks is critical in terms of protecting valuable equipment (lost refrigerant causes excessive wear), because of the cost of lost refrigerant, and in terms of complying with regulations (ASHRAE Standard 15-1994 recommends a vapor detector in each machinery room that will actuate an alarm when leaks are detected, initiate mechanical ventilation, and shut down flames in boilers, etc., in case a refrigerant leak exceeds its threshold limit). Note that ASHRAE standards are only recommendations and it's the local or city agencies that have governing authority; you must comply with local building codes and regulations (which usually embody some version or adaptation of the ASHRAE standard).

Leaks from equipment and pipes can be detected using soap bubbles, dyes, or portable, hand-held electronic devices to pinpoint a leak at some specific system location. Also, leaks can be detected in an area or room using stationary monitors that operate all the time. The leak can then be isolated using a pinpointing device.

There are three main attributes of continuous monitoring detection equipment: sensitivity, selectivity, and cost.

Sensitivity

Detectors with high sensitivity can detect leaks of refrigerant vapor in very small concentrations, while low-sensitivity detectors need higher concentrations of refrigerant. A detector with high sensitivity may be able to accurately discriminate vapor concentration levels of less than one part per million (ppm), while a low-sensitivity detector may be able to discriminate only in increments of 100 ppm or higher. Low-sensitivity detectors are non-selective (see below) and may respond to paints, cleaning fluids, or other non-refrigerant gases.

Sensitivity limits are determined based on the refrigerant safety classifications (A1, B1, etc.) and their respective allowable exposure limits.

One common way to measure how "sensitive" a detector is, is the unit's "detection limit," which is the minimum amount of material it can sense. A high-sensitivity device does not necessarily have a low detection limit although these factors usually correspond. The sensitivity of a device is determined by the detection technology and the material being detected. There are two typical ways to measure detection limits: ounces per year for pinpointing applications and ppm for area monitoring. Portable leak pinpointing equipment have detection limits around 0.25 ounces per year, while area monitors have detection limits as low as one ppm.

Because the sensitivity required can vary with different refrigerants or compounds, detectors must be matched to specific applications. Some manufacturers provide single instruments that have various sensitivity settings, or use multiple technologies to meet various sensitivity requirements.

Selectivity

Selectivity is the ability of an instrument to detect and discriminate only a particular refrigerant while ignoring other compounds that may be present in the area. This feature can be important when multiple refrigerants are present, or if certain paints or cleaning fluids are present.

In terms of selectivity, leak detectors can be categorized as non-selective, halogen-selective, or compound-specific. In general, as the specificity of the monitor increases, so does the complexity and cost.

Non-selective detectors will detect not only refrigerant vapors, but any non-refrigerant gases that may be present regardless of their chemical makeup. As an area monitor, these non-discriminating detectors can sometimes cause false alarms. As a result, non-selective detectors are typically hand-held devices that have limited use for area monitoring. These portable devices are usually very rugged and relatively inexpensive (normally less than $500). These detectors are based on one of several technologies: electrical ionization, thermal conductivity, or metal-oxide semiconductors with detection limits usually between 100 ppm and 500 ppm.

Halogen-selective detectors can either be stationary area monitors or portable devices used to pinpoint specific leaks. Halogen-selective detectors use a specialized ceramic metal oxide semiconductor (CMOS) sensor or a heated diode sensor that specifically detect fluorine or chlorine. They discriminate fluorine or chlorine fairly well from other gases and reduce the number of false alarms. These easy-to-use detectors are durable and have higher sensitivity than non-selective detectors (detection limits are typically 10 to 20 ppm when used as an area monitor). They cost from $1500 to $3000 for single-point detection. Although these sensors are halogen-selective, they are not compound-specific. This may not pose a problem if only one refrigerant is used in the monitored area.

Compound-specific detectors are the newest type. These can detect specific compounds using infrared-based (IR) technology in which specific IR light wavelengths are absorbed by refrigerant molecules and produce a unique, identifiable signature. IR detectors can be very sensitive and normally have detection limits from one to five ppm. Although they are designed to detect specific compounds, they can respond to gases with similar signatures if the concentration is high enough. Costs range from $4000 to $7000.

Some monitors can be set up to switch automatically between gases being monitored. This lets you monitor for a specific gas in areas that contain several different refrigerants. Most have communication interfaces that can be connected to remote terminals that let you monitor system performance and detector status. Detection systems can be expanded to monitor over 30 individual areas and accurately measure a different gas in each area.

Selecting an Area Monitor

The decision to purchase an area monitor involves many factors including (but not limited to):

- Where is the monitor to be located?
- Other than refrigerants, are there other chemicals in the room?
- Will you need a monitor that is compound-specific?
- What level of detection will be needed?
- How many detectors are needed?
- What is your budget?

If the equipment room is protected from outside vapors entering the area, a halogen-specific sensor could probably be used as an area monitor. If the equipment room is exposed to outside vapors that would be detected by a halogen-specific system (as from the ventilation system in a chemical plant), a compound-specific instrument is probably best.

The highly selective IR instruments may be best where the air in a machine room could be contaminated by many types of gases, different refrigerants, solvents, propane, butane or natural gas, or in a situation where it is important to be able to detect different exposure levels (for example, 30 ppm for one substance and 1000 ppm for another).

If the entire equipment room only uses one refrigerant, a halogen-specific detector may do the job. If there are different types of refrigerants present, a halogen-specific detector can work if the exposure levels of the refrigerants are about the same. This choice depends in part on whether the monitor is used only as a safety feature, or for detecting a specific type of refrigerant leak.

As a safety feature, multi-level alarms interlocked with the ventilation system may be activated at specific detection levels. Oxygen monitors are often combined with the system to assure adequate levels of oxygen are present. All detection systems should be able to initiate fans and alarms through communication with building automation systems. Also, all systems should offer the flexibility of multiple detection points or detection of multiple gases, and should be able to respond quickly to an alarm condition.

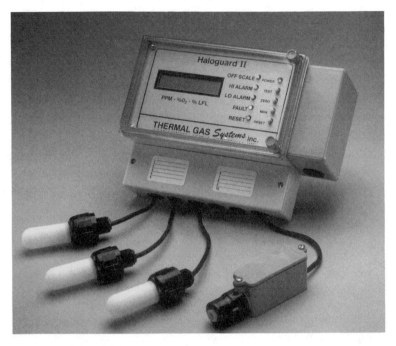

Figure 5-2 Multichannel Multigas Monitor (Thermal Gas Systems Inc.)

Implementing a leak detection system is a rather specialized task—you should consult with detection system experts. Some suppliers of detection equipment are:

- Yokogawa —CMOS and IR; Newnan, GA
- Thermal Gas Systems, Inc.—CMOS and IR; Roswell, GA
- Mine Safety Appliance (MSA)—CMOS and IR; Pittsburgh, PA
- SenTech—Ionization; Indianapolis, IN

"Tighten Up" Equipment to Reduce Refrigerant Losses
You can lose valuable refrigerant through leakage as a result of:

- Charging a system with "new" refrigerant.
- Transferring refrigerant to or from recovery devices.
- Tube leaks and seal problems.
- Lack of proper leak testing.
- Inefficient procedures for purging and for handling and storing refrigerant.

Some steps that you can take to help avoid these losses include:

- Avoid vacuum or positive pressure systems.

    When a low-pressure chiller is idle it creates a vacuum that sucks in moisture-laden air (which has a negative effect on system performance) and causes leaks to develop. A vacuum system pressurizer can help avoid this problem. This permanently mounted service tool consists of a refrigerant pressure controller (accurate to 1/100th of a pound), heat blankets located under the evaporator, and a pressure transducer attached to the condenser gauge line. In some cases, this type of system can pay for itself by reducing refrigerant losses by up to 90%. (Note that a system pressurizer is not necessarily applicable in warm weather climates.)

- Use no-leak or low-leak mechanical shaft seals.

- Implement microprocessor-based controls and a monitoring system

- Control purges and pump-outs (discussed next).

- Use low-loss fittings for hoses in portable equipment when reclaiming and recycling.

Control Purges and Pump-Outs

Low-pressure chillers must periodically be purged to remove the air brought in by the vacuum or low pressure they create. Inevitably, some refrigerant vapor will be purged along with the air removed from the system. Many older purge devices allow three to eight pounds of refrigerant to be removed with each pound of air purged from the system.

New purge devices have been developed with refrigerant conservation in mind, and some are virtually emission-free with efficiencies as high as 99%. Although high-efficiency purge units typically cost between $4500 and $5500 per unit, they can be paid for quickly. Some systems include a counter that indicates if there is excessive purging when the chiller is running. (This would indicate that it's time to check the chiller for leaks.) In addition, there are purge-recycle devices on the market that let you capture, store, and filter the refrigerant that escapes with the purged air.

Chillers that hold refrigerant charges of several pounds must also be equipped with pump-out units. These are closed-loop systems with a pump and storage tank to hold refrigerant while the chiller is being serviced. Low-pressure chillers also may need a permanent pump-out unit that is used to remove the refrigerant at times when the chiller will stand idle for a long time (as in the winter if air conditioning is not used). This is done to make sure the system does not collect non-condensable substances as it sits idle.

The EPA rule regarding recovery and recycling specifies pump-out evacuation levels (in inches of mercury) for low-, medium-, and high-pressure systems.

Some specific steps that you can take to avoid excess refrigerant loss during purges and pump-outs include:

- Install high-efficiency purges (especially for R-11 machines).
- Change flair fittings to brazed.
- Install back-up relief valves.

Avoid Catastrophic Failures

A catastrophic failure is when all or most of the refrigerant in a system is accidentally released to the atmosphere. Relief-pressure devices are designed to help avoid permanent damage to a chiller as a result of an over-pressure emergency.

Rupture discs are the kind of relief-pressure device used in most large-tonnage CFC chillers. Once an over-pressure emergency engages this device, it will release the *full* refrigerant charge safely into the atmosphere.

Spring-relief valves are typically a better alternative for relieving pressure. Unlike a rupture disc that does not close the refrigerant line after it ruptures, this type vents off only enough refrigerant to lower the system to a safe pressure and returns to a closed position as soon as the system pressure returns to normal.

A rupture disc can be used in series with a spring-relief valve. If the chiller's disc ruptures, the carbon elements of the disc break apart and are trapped in a chamber, and the refrigerant flow is diverted to the spring valve where relief gas is vented to an outside vent line. An advantage of this relief system is that the refrigerant charge does not need to be removed to install it. (Non-fragmenting flexible discs are also available.)

Recover, Recycle, and Reclaim

Maintaining equipment with recycled or reclaimed refrigerants may be a viable option if you have a large inventory of equipment. This would let you gradually retire some units, reclaiming or recycling the refrigerant to keep other units going and put this valuable resource to good use.

Section 608 of the Clean Air Act governs various aspects of recovery and recycling. The essence of this section are the regulations that require certified refrigerant recovery and recycling equipment to prevent refrigerant venting during repairs, maintenance or servicing, charging or removal of refrigerants, and when old equipment is discarded.

Also, the Clean Air Act requires that you keep current all the required paperwork (records of refrigerant purchases, recovery equipment, technician training, refrigerant leaks, and disposition of recovered refrigerant, etc.). Section 608 also sets standards for reclaimed refrigerant if it is sold to a refrigerant reclaiming business. (Under current regulations, used refrigerant cannot be sold to other users. An owner can move used refrigerant between facilities at will. Used refrigerant must be sold to a certified reclaimer, who can then purify and resell the refrigerant.)

In order to either recycle or reclaim refrigerant, it must be recovered—that is, removed and contained. This should be done using EPA-approved recovery or recovery/recycle equipment. Some recovery systems simply remove the refrigerant from existing equipment and contain it for later clean-up. Other equipment has recycling capabilities built in to the recovery system. Storage and transport vessels, additional parts of the recovery process, also must meet EPA requirements.

As mentioned earlier, recycling refrigerant refers to doing "basic clean-up" on used refrigerant so it can be used by other equipment in your facility. Reclamation refers to performing a more rigorous clean-up so the refrigerant will meet certain standards and can be used in equipment that others own. The recycling process involves cleaning recovered refrigerant by means of oil separators and filter-dryers.

There are potential problems with using recycled refrigerants. Often, technicians are often not aware of the potential problems caused by recovering different types of refrigerants with the same recovery device, which means you cannot be certain that all contaminants have been removed from the system and refrigerant purity cannot be assured. (Technician certification is now required by Section 608.) Contamination will occur if systems are recharged with recycled or recovered refrigerant from contaminated cylinders and recovery machines. Recycle equipment is only as effective as the filter-dryers or oil separators. Filters must be changed between recovery/recycle projects to avoid cross contamination.

Passive recovery is often used to capture refrigerant from small appliances. This approach, sometimes called "system-dependent" recovery, uses the appliance's compressor to pump out the refrigerant into a recovery cylinder.

Another passive recovery system used by many businesses that repair appliances uses a specially built plastic bag to recover refrigerant. The leak-proof bag is placed around an exit port (or valve) and collects the refrigerant as it is pumped out by the compressor.

Another type of passive recovery system uses a "cold-recovery" technique which is used in the case where the compressor has failed and cannot pump out the refrigerant. A line is attached from a system refrigerant line to a recovery cylinder. This cylinder is placed in ice which causes it to become colder than the refrigerant loop. This difference in temperature drives the refrigerant out of the system lines into the recovery cylinder. The refrigerant is collected in the cylinder as it condenses into a liquid.

An *active recovery* system has a dedicated compressor in the recovery unit itself. This "self-contained recovery" approach is the most common, and is used with all high- or very high-pressure equipment. There are a wide range of active recovery units on the market. Recovery equipment for high-pressure equipment of 25 tons or less costs around $1000. Recovery equipment for larger high-pressure units, with up to 1,000 pounds of refrigerant, may be as much as $8000.

Some specific steps you can take if you plan to recover and reclaim or recycle refrigerant include:

• Carefully review the EPA (and any local) regulations governing recovery, recycling, and reclaiming of refrigerants.

• Buy or rent a portable or on-site storage and recovery unit to help keep your costs down.

• Inspect the refrigerant access servicing and recovery apertures to ensure they are not leaking.

- Inspect the recovery or storage system's piping and hoses to make sure they are not causing unnecessary refrigerant loss.

- Put in place an appropriate stockpiling or banking plan for the refrigerant you recover.

Stockpile or Bank

The EPA suggests that companies may want to "bank" CFC refrigerants, but they do not have much data on how much is actually being stored as old equipment is replaced. It is safe to assume, however, that many building owners and managers are holding on to their R-11 and R-12.

If you elect to stockpile refrigerant, you will need to keep a detailed inventory of what you have and where it is stored. Also, since CFCs are an increasingly valuable commodity, you will need to think about security—where can you store it so that it will not get "lost"? In addition, you will need to understand any local codes, such as fire regulations, that can have an impact on where and how you store your refrigerant. Some refrigerant wholesalers have banking programs that will help off-load some of the details associated with keeping a store of CFCs.

In situations where new non-CFC equipment has been added to facilities, recharging systems becomes more complex and can present potential problems. Technicians must be properly trained to correctly identify refrigerants so that systems are not charged with the wrong refrigerant. A chiller that has a mixture of CFCs and non-CFCs can cause major problems.

Retrofit or Conversion

If your plan includes retrofitting or converting either a chiller or a commercial refrigeration system, there are several issues to consider.

Retrofit and Conversion Alternatives

As mentioned earlier, retrofit and conversion options range from those that are relatively simple and inexpensive to more complex and expensive approaches.

- **Simple retrofits** can be as basic as just changing the equipment's refrigerant and oil. In general, this approach is the least expensive, but can create significant "penalties" in terms of decreased chiller efficiency.

- **Optimized conversions** include changing the gears in the equipment to tune it for operating with a new refrigerant. While this approach is more expensive than a simple retrofit, it usually offers better chiller performance.

- **Driveline retrofits** involve replacing the motor and compressor driveline assembly and installing new microprocessor controls. Although this is the most expensive retrofit/conversion alternative, a properly engineered driveline retrofit can virtually eliminate any degradation due to using a new refrigerant.

Retrofitting or Converting Chillers

According to manufacturers, the number of chiller conversions doubled or tripled between 1992 and 1993, and doubled again from 1994 to 1995. However, this does not represent the vast numbers of conversions you might think: fewer than 400 chillers were converted before 1993.

This means that the anticipated rapid growth will still result in relatively few converted systems—probably between 1,600 to 2,400 out of an installed base of 80,000 or more CFC chillers.

In part because of the relatively limited base of converted systems, many people think that retrofitting or conversion is difficult and unreliable. This is not necessarily the case—particularly if you carefully determine which systems you want to retrofit or convert, and know when the changes are to be made and how much you can expect to pay.

Chillers that are between two to ten years old often make good retrofit/conversion candidates because they usually are designed for improved efficiency and typically have a long life ahead of them. In addition, these newer chillers often can be retrofitted or converted using relatively simple, less-expensive techniques.

When considering a retrofit or conversion of an older chiller (from 10 to 20 years old) you need to assess carefully the tradeoff between the complexity and cost of the conversion and the operating efficiency after the conversion. For example you might want to consider:

- Changing the driveline and adding a high-efficiency purge. (This would require that you install a refrigerant sensor to meet Standard 15 provisions.)

This alternative is relatively complex and expensive. However, it offers the advantages of little or no degradation in system performance.

- Changing from R-12 to R-134a—simply a change in refrigerant and oil. (This would require that you install an oxygen-depletion sensor to meet Standard 15 provisions.)

This alternative is comparatively simple and less expensive, but generally results in a lower chiller cooling capacity. However, this may not be a serious drawback if you don't need to run the system at its full capacity.

Most original equipment manufacturers can give you information on special design requirements (such as material compatibility and lubrication) and they may be able to help you to analyze the energy and capacity tradeoffs of retrofitting. You also may want to enlist the assistance of a design professional who can use the estimated cost and efficiency numbers to do a computerized life-cycle cost analysis. This can help you make a sound financial decision based on simple payback, internal rate of returns, cash flow, or other financial data for the various options.

When to Retrofit or Convert A Chiller

One way to keep down the costs of retrofitting or converting a chiller is to schedule it at the same time as standard servicing or, ideally, during a major system overhaul (which is recommended every seven to ten years.)

In fact, some industry sources indicate that you could expect to save between $10,000 and $15,000 on the conversion of a typical 500-ton hermetic centrifugal chiller by scheduling the conversion during a major overhaul. Some of the reasons that this approach can save so much money include:

- When seals and gaskets are being changed anyway, it is comparatively easy and inexpensive to replace them with seals and gaskets compatible with both traditional and alternative refrigerants.
- During a major overhaul, the motor typically is removed and sent out for inspection and rewinding if necessary. Why not replace a hermetic motor with one compatible with new refrigerants—or with a retrofitted motor (like the one described later)?
- Since the machine is already open, it is comparatively easy to replace the gear-impeller sets for gear-drive machines or modify the impellers for direct-drive machines.

The Motor Retrofit Option for Chillers

All electric chillers have motors to drive the system, and a motor retrofit may be a viable option to other methods of CFC chiller conversion. Unlike motor rewinds, that typically result in a loss of chiller efficiency, a motor retrofit is a re-engineered stator-rotor replacement that includes upgraded insulation so you can use additional stator metal and rotor windings.

This approach typically results in little or no loss of efficiency. In fact, according to MagneTek, Inc., a vendor of motor retrofits, this approach can actually boost efficiency by 3%.

This approach can offer several benefits in addition to improved operating efficiency. Specifically, these retrofitted motors are said to run cooler, have more torque, and have a longer life. In addition, the motor conversion may be done in as little as 48 hours during a typical one- to four-week engineered chiller conversion process.

Chiller Retrofit and Conversion Costs

A chiller retrofit or conversion can cost anywhere from 20% to 80% of the total installed cost of a replacement chiller, depending upon the complexity of the method you choose.

An "average" conversion would cost about 50% of the installed cost of a new chiller, but generalizations are difficult to make because the cost for a particular retrofit depends on the chiller model to be reworked and the details of the installation.[39]

The graphic in Figure 5-3 (from P. John Ostman; *ASHRAE Journal,* May 1993) shows these relative costs.

The cost of converting a chiller can vary widely. To convert a medium-capacity open-drive chiller will cost between $30,000 and $50,000. Larger chillers with hermetic compressors are even more expensive to convert. Including the required motor change, converting a 300-ton unit may cost as much as $60,000, and a 500-ton unit as much as $75,000.

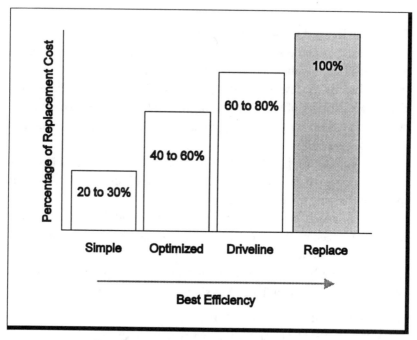

Figure 5-3 Chiller Retrofit and Conversion Costs

Keeping in mind that actual conversion costs can vary dramatically, useful rules of thumb for estimating costs are:

- $60 to $140 per ton to convert from R-11 to R-123.
- $150 to $200 per ton for a new chiller (equipment only).
- $275 to $400 per ton for the total installed cost of a new chiller. (This is typically about 1.5 to 2 times the cost of equipment.)

When considering the above rough cost guidelines, you also should keep in mind that a new chiller may be able to pay back its higher first cost in an extremely short time simply through reduced energy costs.

Retrofitting or Converting Refrigeration Systems

Although the life of refrigeration systems can be up to 20 years, supermarkets are typically remodeled or undergo reconstruction every seven years or so. This is an ideal time to consider the condition of the existing equipment and to determine if a system retrofit or conversion makes sense according to your plan.

For refrigeration systems, there will be a much greater range of possible solutions in terms of refrigerant and equipment options when compared with air-conditioning equipment.

There are two primary conversion options: retrofit the system to use short-term (or interim) refrigerants that will be phased out sooner rather than later (such as R-22, and blends that contain HCFCs and therefore chlorine), or retrofit the system to use long-term refrigerant alternatives that do not contain chlorine (such as pure HFCs or blends that contain HFCs.) For newer equipment, it may make sense to convert to one of the alternative refrigerants.

Refrigeration System Retrofit and Conversion Costs

The capital requirements to convert refrigeration systems also will vary depending on the situation.

Conversion of conventional systems to interim HCFC refrigerants will cost approximately $12,000 to $15,000 for average stores, and up to $30,000 for large stores with conventional systems; not including refrigerant.

Using our previous medium-sized typical store chain, the table below shows the low and high extremes that retrofit costs may fall between.

This shows a very wide range of possible implementation costs and associated operating costs.

Typical Medium-Sized Store Refrigeration System Conversion Costs				
Retrofit Choice	Refrigerant	System	Implementation	
			Cost / Store	Total Cost
Minimum first cost	Interim HCFCs	Existing	$12,000	$1,200,000
Minimum operating cost	Zero Cl HFCs	New Parallel	$250,000	$25,000,000

Table 5-5 Typical Medium-Sized Store Refrigeration System Conversion Costs

New Chiller Equipment

Aging chillers (15 to 20 years old or older) and other equipment nearing the end of their useful lives are good candidates for replacement with new, non-CFC units.

If you will need or want to replace a chiller in the near- to mid-term future, you should be sure to plan and schedule it well in advance. There recently has been a significant increase in the sale of replacement chillers. For example, in 1993:

- Shipments of new, large centrifugal chillers set a record high of almost 6,000 units (according to ARI, this is an increase of over 23%).

- Shipments of new reciprocating liquid chillers also set a new record (9% higher than the previous year).

As sales of new equipment increase, expect delivery to take longer. Even streamlined manufacturing procedures and reduced assembly-line times can not accommodate such rapid growth without delays in delivery. The major chiller manufacturers have indicated that they do not plan to increase their production facilities simply to accommodate a frenzy of short-term buying. A representative from one major manufacturer reports that some major companies are asking for guaranteed future production slots.

As we have said earlier, older chillers were quite often sized using estimated load data and do not accurately match the chiller capacity with the actual air-conditioning load for the building. Also, the load over the years may have changed dramatically. Equipment replacement offers an opportunity to correct this mismatch. A careful analysis can determine more accurate chiller load requirements, which will result in higher operating efficiencies (and most likely lower operating costs) and better system performance.

Chiller Replacement Cost Considerations

As discussed above, replacement chillers typically have significantly higher first costs than a retrofit or conversion—on average, centrifugal chillers cost from $250 to $350 per ton.

However, the capital costs associated with buying and installing a new chiller often can be offset by reduced operating costs due to improved efficiencies. For example, consider a water-cooled centrifugal chiller:

- In the 1970s, an efficient unit of this type would have efficiencies around 0.8 kW/ton to 1 kW/ton.

- Today's more efficient units of this type have efficiencies from 0.50 kW/ton to 0.65 kW/ton.

Chiller replacement may be the best option if the chiller is at the end of its life. In addition to improved efficiencies (new chillers are tighter), you can typically count on savings in terms of reduced maintenance requirements when you replace an aging chiller with a new one.

Even with a five- to seven-year payback period, the entire replacement chiller installation can be financed, so that the monthly interest and principal payments will cost less than the monthly energy savings.

When selecting a replacement chiller system, you should analyze carefully your cooling load requirements. By selecting a high-efficiency chiller, and making sure it is sized correctly for actual loads, you often can significantly reduce the amount of refrigerant you need to support the system.

Also consider operations and maintenance information, which will help you determine the amount of redundancy you should have in order to provide the level of reliability you want.

If your facility has multiple buildings or one building with multiple chillers, you may want to consider replacing these chillers with one central chilled-water system. This approach would let you reduce the total amount of required capacity by taking advantage of the diversity among the different buildings or areas in one building and reducing the amount of redundancy. This approach may let you reduce the total connected chiller tonnage by 20% to 35%—or in some cases up to 45%.

New Refrigeration Equipment

The decision to install new commercial refrigeration equipment is based on many factors. As discussed in Chapter 4, the condition of the existing equipment and the economics of installing new equipment play an important role in the decision.

If the existing equipment is at or near the end of its useful life, the system is probably not operating efficiently. So, for stores or facilities with good future potential, instead of spending money on conversion or improved containment of the old systems, investing in a new system may be the best alternative.

Obviously, any system that is unreliable or experiencing a high CFC leak rate which requires a large annual refrigerant charge is a good candidate for a change—either a system retrofit or replacement with new equipment. Also, a system where the condenser or other major component is in need of replacement is a likely candidate. If there is any equipment that would otherwise need to be replaced, the economics of the decision is improved. It's better to contribute to a new system and reap a "cost avoidance" of replacing the faulty component on an older system. In addition, if the refrigeration equipment is fully depreciated, you will avoid writing off existing assets.

In general, a store that is expected to stay in service for 15 years or more and that is anticipating remodeling is a good candidate for new equipment, especially a relatively large store where the energy savings of a new system are greater. A new system configuration can be sized to allow flexibility for future expansion and increased loads.

If you decide to buy new refrigeration equipment to replace existing equipment, two viable options are to install new equipment that uses R-22, or that uses a long-term HFC refrigerant. And, a parallel system design that uses one of the new, long-term refrigerants offers the best overall economic and environmental choice. This type of system may require more up-front expense, but because it is much more energy efficient, energy costs can be reduced up to 40% which can contribute to paying for the new equipment in five to eight years.

New equipment replacement costs can be high: for a typical 30,000 to 40,000 square foot store, to implement a new parallel, state-of-the art system using zero-chlorine refrigerant options may cost as much as $220,000 to $260,000 per store.

However, annual energy cost savings can be in the range of $45,000 to $50,000. At this rate, simple payback calculations range from 4.5 to 6 years.

MONITOR AND EVALUATE

Even the best laid plans will require some modification over time, as technologies change, regulations are updated, and unanticipated facility needs occur.

Monitoring and evaluation is an ongoing process in which you track your actual progress against your plan, and revise your plan according to the external realities you are dealing with.

It is generally a good idea to summarize any significant changes in writing, and make sure you communicate these to all persons who are involved with executing the plan—from upper-level management to staff personnel who are directly responsible for carrying out the day-to-day operational and maintenance activities.

UNISYS: AN EXAMPLE PLAN

The following summary of the Unisys refrigerant management plan is from a report to the 1994 Globalcon by Ellen Miclette, Senior Marketing Engineer, The Trane Company.

Unisys is a $8 billion company that provides information systems and services to their customers in more than 100 countries. They have assigned a VP of Corporate Environmental Services who is on the same level in the organization chart as the VP of Facilities Operations and Management. Both of these positions report directly to the CFO and then to the CEO. This demonstrates how important environmental management is to Unisys.

The VP of Corporate Environmental Services represents centralized management and control of CFCs and energy use. This person controls the budgets to implement corrective action and compliance strategies for all Unisys U.S. locations.

Unisys began their CFC management program around January 1, 1994, and a member of the facilities management team outlined the requirements for removing CFCs from their physical plants. Recognition of the need and support from corporate facilities executives was essential for proceeding with the plan.

At the time of the study, Trane considered 18 sites with a total of 68 installed chillers. Most of these chillers were designed for low-pressure refrigerants, and ranged in size from 100 to 1000 tons.

Unisys enlisted the assistance of The Trane Company to help fill in the details of the plan, outlining the right facts and the best arguments that would help them sell the plan to executive management. The gradual process began with an effort to increase awareness of Freon issues, then to overview existing inventory, and finally to understanding the economic opportunities for equipment upgrades.

Ms. Miclette reports that the Unisys plan had the following five components:

- CFC Policy Development.
- Immediate Action Plan.
- Asset Assessment.
- Conversion/Replacement Plan.
- Service Policy Development.

CFC Policy Development

According to Ms. Miclette, Unisys recognized the need for a comprehensive policy or mission statement, that they could use to guide them as they created their refrigerant management strategy. Their policy/mission statement is as follows:

As owners and operators of commercial refrigeration and air conditioning systems, we will make every effort to eliminate the release of chlorofluorocarbon containing compounds into the atmosphere. Our efforts will include, but are not limited to the following:

1. *Rigorous compliance with federal, state and local statutes governing the management of CFCs, HCFCs and any other substance that may be regulated by EPA.*

2. *Absolute conformance to practices and procedures designed by the air conditioning and refrigeration industry in order to mitigate and to control refrigerant emissions.*

3. *Training and certifying our personnel in refrigerant handling and management techniques.*

4. *Instituting refrigerant management elements in our internal quality assurance programs.*

5. *Directing and supervising the efforts of service contractors to ensure their compliance with our refrigerant management plans.*

6. *Expanding our refrigerant management practices to include Halon and other CFC-containing compounds as required.*

In the interest of minimizing the global environmental harm associated with CFC releases, we will make refrigerant management a very high priority for all operations and maintenance personnel. We believe that these measures will lead to a more responsible treatment of CFC containing compounds throughout our portfolio of managed properties.

Immediate Action Plan

The next part of the plan that Unisys developed outlined the steps that should be taken immediately to comply with current EPA containment regulations.

This included equipment evaluation for retrofits, and consideration of high-efficiency purges, pressurizing rupture disc protection, and refrigerant cleaning devices. It also included refrigerant recovery equipment and equipment room safety.

Asset Assessment

In conjunction with the immediate action activities, in the Chiller Asset Assessment part of their five-part plan, Unisys collected data on all of their existing facilities. The data they collected basically covered the items that were recommended earlier in these materials.

Conversion/Replacement Plan

To develop the conversion/replacement plan they evaluated the equipment keeping in mind the choices that are available regarding what to do with the existing equipment. Ms. Miclette reports that Unisys considered the following options:

- Containment of refrigerant per immediate action plan.
- Scrap the equipment.
- Convert and optimize the equipment to R-123.
- Retrofit a new compressor optimized to R-123.
- Replace with new, like equipment or new, unlike equipment (e.g., gas cooling, air-cooled, different capacity chillers, or thermal energy storage equipment.)

For the purpose of making the decisions for the equipment, Unisys appraised each unit on such criteria as the current and future requirements of the system, machine age, condition, energy efficiency, predicted maintenance requirements, utility incentives, and first costs for all options.

By demonstrating that retrofitted and/or new equipment would result in vastly improved average kW/ton and Delta, the plan team was able to convince the financial managers that their plan had merit. With just the efficiency improvements alone, and using conservative numbers, at a minimum Unisys could realize an annual savings of about $1 million. And the savings could be further increased when combined with plans to further improve total system efficiencies by implementing improved controls and building automation, variable frequency drives, improved hydronic system configurations, and demand-shifting strategies.

The preliminary evaluation determined that of the 68 installed chillers considered for the study, 27 will be converted to R-123, 34 will be replaced, and seven will be scrapped. Unisys will also "bank" the reclaimed refrigerant from the converted, replaced or scrapped chillers to use with the remaining CFC-based chillers.

Service Policy Development

Finally, Unisys implemented a policy for all service—both in-house servicing and that done by outside contractors. Service guidelines were based closely on their policy statement; and, as Ms. Miclette reported, it consisted of the following items:

- All personnel will maintain and follow a refrigerant management policy that includes a goal of zero refrigerant emissions.
- All technicians will be trained and certified on proper refrigerant handling.
- All personnel will use proper recovery and recycling equipment and techniques.
- Records will be kept as required by EPA of all refrigerant usage.
- Refrigerant and oil will be disposed of according to national and individual state regulations.
- All personnel will understand and follow Clean Air Act provisions.

Ms. Miclette reports that Unisys has already started implementing their plan and seven plants have replaced and converted equipment and implemented system upgrades.

CHAPTER SUMMARY

A plan to manage refrigerants helps you determine the best course of action to take as you implement the transition from CFCs to environmentally acceptable refrigerants.

The planning effort requires researching and analyzing your situation to understand as best you can what your alternatives are.

There are many variables among different facilities, equipment, personnel, and other requirements. You must develop your own approach—one that is tailored to your specific situation and based on an individualized analysis and evaluation.

A sound plan will serve as the basis for making crucial decisions. The five main steps to a sound approach include:

- **Organize for Action**—put together the people and communication systems you will need.

- **Analyze the Situation**—gather data about your current situation, including your equipment and constraints.

- **Formulate an Action Plan**—evaluate your alternatives to determine the specific actions you should take, and when you should take them.

- **Execute the Plan**—proceed to implement improved maintenance and containment procedures, conduct conversions and retrofits, and purchase and install new equipment as outlined in the plan.

- **Monitor and Evaluate**—keep an eye on your progress and any changes (available technologies and regulations) that can have an impact on your plan to make sure you stay on track.

In the course of the planning process, you'll consider many alternatives—and the number of alternatives to consider will be in direct proportion to the amount of cooling or refrigeration equipment involved.

Many choices are made—some out of principle, most for economic reasons. Successful planning results in choosing the right combination of options at the best possible time and leads to a wise use of resources. The final outcome is prudent operation and a healthier planet.

Glossary

ANSI: American National Standards Institute.

ARI: Air-Conditioning and Refrigeration Institute.

ASHRAE: American Society of Heating, Refrigerating, and Air-Conditioning Engineers.

ASME: American Society of Mechanical Engineers.

ASTM: American Society for Testing and Materials.

Azeotrope: A refrigerant blend that behaves like a single fluid in that it maintains a constant saturation temperature during evaporation or condensation. (See Zeotrope.)

Blends (Mixtures): Refrigerant blends are combinations of two (binary), three (ternary) or more chemical components. For example, R-500 is a binary blend of 73.8% CFC-122 and 26.5% HFC-152a.

CFCs: Chlorofluorocarbons are compounds that consist of chlorine, fluorine and carbon atoms. They contribute to the depletion of the ozone layer and may remain undiminished in the troposphere for 100 years or longer.

Energy efficiency: The ratio of the work output to the work input.

EPA: Environmental Protection Agency.

Flammability: The relative ability of a refrigerant to ignite. It is dependent on many factors, including the refrigerant to air ratio, pressure, temperature, and ignition source. There are many standards for measuring flammability.

Fluorocarbons: Chemical compounds that include CFCs, HCFCs, and HFCs.

Global warming: An increase in the natural greenhouse effect of the Earth which leads to the increased heating of the earth. Solar energy is absorbed by the Earth's surface and some of the resulting heat energy is radiated back into the atmosphere. Increased concentrations of man-made gasses, including carbon dioxide, methane, and CFCs absorb and trap the radiated heat warming the Earth's surface.

Global Warming Potential (GWP): The relative ability of a green-house gas to trap heat in the atmosphere. It is measured relative to either CFC-11 (Halogen GWP) or carbon dioxide. GWP can be either direct, from the actions of refrigerants, or indirect, from the carbon dioxide produced at power plants when making electricity.

HCFCs: Hydrochlorofluorocarbons are chemical compounds that consist of hydrogen, chlorine, fluorine and carbon atoms. These have many of the useful properties of CFCs but 90 to 98% are destroyed naturally in the lower atmosphere. This makes their ODP considerably less than CFCs.

HFCs: Hydrofluorocarbons are chemical compounds that consist of hydrogen, fluorine and carbon atoms, which like the HCFCs, are natu-rally destroyed in the lower atmosphere. Because they have no chlorine, they do not deplete the ozone layer.

Hygroscopic: Capable of absorbing and retaining, or losing moisture.

Miscibility: The ability of a liquid or gas to dissolve uniformly in another liquid or gas.

Ozone Depletion Potential (ODP): The relative ability of a substance to destroy stratospheric ozone, which protects the earth from harmful ultraviolet radiation. It is measured relative to CFC-11.

Reclaim: Refrigerant is removed from refrigeration or air conditioning systems and is chemically reprocessed, usually at reprocessing or manu-facturing facilities, for further use. All reclaimed refrigerant must be verified as meeting ARI Standard 700 levels of purity.

Recover: Refrigerant in any condition is removed from refrigeration or air conditioning systems, is then stored, reclaimed, transported or recycled. Testing or processing is not required.

Recycle: Refrigerant is cleaned for reuse, but does not meet reclaimed standards. The amount of contaminants are reduced by oil separation, removing non-condensable agents, or by single- or multiple-pass processing. Moisture, acidity and particulate matter are reduced. Recycling is usually done at the field job site or at service shops.

SNAP: The EPA's Significant New Alternatives Policy, part of the Clean Air Act.

TEWI, Total Equivalent Warming Impact: The sum of the direct (chemical) and indirect (energy-related) emissions of greenhouse gasses.

UL: Underwriters Laboratories, Inc.

Zeotrope: A blend of refrigerants that do not behave like a single fluid in that they can be separated by evaporation. (See Azeotrope.)

Resources/Credits

1. EPA, "Regulatory Impact Analysis, The National Recycling and Emission Reduction Program, (Section 608 of the Clean Air Act Amendments of 1990)," Prepared by ICF Incorporated, Prepared for Stratospheric Protection Division, United States Environmental Protection Agency, March 25, 1993.

2. James G. Crawford, James M. Calm, P.E., *ASHRAE Journal*, January, 1994, Industry News, "CFC Conference Attracts Worldwide Attention," (citing Sherwood Rowland, University of California, Irvine).

3. TRANE, CFC Update, Volume 18, January, 1995, James Wolf, Chairman Alliance For Responsible Atmospheric Policy quoted.

4. *Air Conditioning, Heating & Refrigeration News,* April, 1994.

5. *Air Conditioning, Heating & Refrigeration News*, January 9, 1995, Editorial Page, "The CFC footnote provides lessons for future action," (Mark Schoeberl cited).

6. *Register,* April 18, 1995, Health and Science, "Arizona will be only state to allow Freon production in the U.S."

   Air Conditioning, Heating & Refrigeration News, April 24, 1995, "Arizona leads CFC rebellion."

7. *Newsweek,* May 16, 1994, "New Hope for Ozone Layer?"

8. EPA, "Listed Substances" table (Ozone Depletion Potential (ODP) of Class I Substances) United States Environmental Protection Agency, March, 1994.

9. *Industry News,* May 1995 "Global Climate Change from CO_2 and CFCs," Matt Chwalowski of the Edison Electric Institute, Washington DC quoted.

10. Steven K. Fischer, Philip D. Fairchild, P.E., and Patrick J. Hughes, P.E., *ASHRAE Journal,* April 1992, "Global warming implications of replacing CFCs"

11. S. K. Fischer, J. J. Tomlinson, P. J. Hughes, "Energy and Global Warming Impacts of Not-in-Kind and Next Generation CFC and HCFC Alternatives," Oak Ridge National Laboratory, Oak Ridge, Tennessee, USA; Project sponsored by Alternative Fluorocarbons Environmental Acceptability Study (AFEAS) and the U.S. Department of Energy; 1994.

12. Fischer, Tomlinson, Hughes, "Energy and Global Warming Impacts..."

13. S.K. Fischer, P.J. Hughes, P. D. Fairchild, Oak Ridge National Laboratory; C. L. Kusik, J.T. Dieckmann, E. M. McMahon, N. Hobday, Arthur D. Little, Inc., "Energy and Global Warming Impacts of CFC Alternative Technologies;" Sponsored by the Alternative Fluorocarbons Environmental Acceptability Study (AFEAS) and the U.S. Department of Energy (DOE), December 1991.

14. *CFC Update,* Volume 18, January 1995, published by TRANE, copywrite American Standard Inc. 1995.

15. EPA, "Background Data on New Regulations."

16. DuPont, U.S. Regulatory Update, January 1994, "Safe Alternatives."

17. DuPont, U.S. Regulatory Update, January 1994, "CFC/Halon Excise Taxes."

18. DuPont, U.S. Regulatory Update, January 1994, "CFC/Halon Excise Taxes."

19. *The Air Conditioning, Heating and Refrigeration News,* October 31, 1994, "Feds Organize Crackdown on Illegal Imports of CFCs."

20. *The Air Conditioning, Heating and Refrigeration News,* February 6, 1995, "Recycled or not, imported refrigerants face excise tax" per EPA.

21. DuPont, U.S. Regulatory Update, January 1994.

22. Charles R. Miro, ASHRAE Issues Manager, and J.E. Cox, P.E., Ph.D., ASHRAE Director of Government Affairs, *ASHRAE Journal,* January 1994, Washington Report, "Clinton Administration's Climate Change Plan."

23. EPA, U.S. Regulatory Update January, 1994.

24. Charles R. Miro, Associate Government Affairs Directors, and J.E. Cox, P.E., Ph.D., ASHRAE Director of Government Affairs, *ASHRAE Journal,* September 1994, Washington Report "Montreal Protocol Assessment in Nairobi, Protecting the stratospheric ozone layer discussed by the plan's signatory nations."

25. EPA, "Stratospheric Ozone Protection Fact Sheet, Temporary Extension of the Section 608 Refrigerant Reclamation Requirements," April 25, 1995.

26. B. Checket-Hanks, *The Air Conditioning, Heating and Refrigeration News,* April 24, 1995, "'Lost' Techs May Miss Deadline For Recycling Certification."

27. EPA.

28. N. D. Smith, Ph.D., K. Ratanaphruks, M.W. Tufts and A. S. Ng, "R-245ca: A potential far-term alternative for ER-11," *ASHRAE Journal,* February 1993.

29. EPA, Air and Radiation Stratospheric Protection Division, March 1994, "Qs & A's On Ozone-Depleting Refrigerants and Their Substitutes."

30. "Energy and Global Warming Impacts of Not-In-Kind and Next Generation CFC and HCFC Alternatives" Prepared by Oak Ridge National Laboratory, Oak Ridge, Tennessee, by Alternative Fluorocarbons Environmental Acceptability Study (AFEAS), and the U.S. Department of Energy, 1994.

31. Excerpts from ARI letter to Mr. Mukesh K. Khattar, Manager, HVAC, Refrigeration and Thermal Storage, Electric Power Research Institute, Palo Alto, CA, from David S. Godwin, Engineer, Research Projects, Air-Conditioning & Refrigeration Institute, Arlington, Virginia, March 20, 1995

32. Earl B. Muir, Senior Vice President, Engineering and Research, Copeland Corporation, *Copeland Environments,* Premier Issue, 1994, "R-22 Replacements Bring Us One Step Closer to a Chlorine-free Environment."

33. Carrier, United Technologies, "Technological Leadership."

34. *Air Conditioning, Heating, & Refrigeration News,* December 12, 1994, "AlliedSignal praises efficiency of AZ-20 as an R-22 replacement."

35. *CFC Update,* from the Electric Power Research Institute's Commercial Building Air-Conditioning Center, "Status of R-22 Alternatives for Unitary HVAC", Issue 10, July 1994.

36. ARI, "Results for R-502 Alternatives" from ARI's Alternative Refrigerants Evaluation Program (AREP).

37. *ASHRAE Journal,* January 1994, "FICs Investigated as CFC Alternatives."

38. *ASHRAE Journal,* April 1995, "ASHRAE '95: Winter Meeting and Show Report, Refrigerant Makers Gear Up Production."

39. P. John Ostman, director of marketing for York International Corp., "Environmental solutions for today's refrigerant challenges," *ASHRAE Journal,* May 1993. Under heading "Chiller Retrofit and Conversion Costs."

OTHER SOURCES

David Wylie, ASW Engineering, Doug Scott, VaCom Technologies, "Refrigerant Choices for Commercial Refrigeration."

"Practical Concerns for Refrigerant Management in Buildings," as presented by David Wylie, P.E., ASW Engineering, Tustin, California, for Globalcon '94 Conference.

"The Retrofit Experience of Unisys Corporation," as presented by Ellen Miclette, Senior Marketing Engineer, The Trane Company, for Globalcon '94 Conference.

"Planning for a Non-CFC Future, Developing a Phaseout Strategy," Norman E. Miller, Associate, ZBA, Inc. Engineers Architects, Cincinnati, Ohio and Mark Spurr, Principal, Resource Efficiency, Inc., St. Paul, Minn., *Consulting-Specifying Engineer,* January, 1994.

ASHRAE Standard 15-1992, Safety Code for Mechanical Refrigeration. ASHRAE, 1992, Atlanta, Georgia.

ANSI/ASHRAE Standard 34-1992, Number Designation and Safety Classification of Refrigerants. ASHRAE, 1992, Atlanta, Georgia.

EPA Stratospheric Ozone Protection Action Guide, United States Environmental Protection Agency, February, 1993.

EPA Stratospheric Ozone Protection Fact Sheet, United States Environmental Protection Agency, March, 1994.

EPA Stratospheric Ozone Protection Final Rule Summary, United States Environmental Protection Agency, June, 1993.

SCAQMD Rule 1415: Reduction of Chlorofluorocarbon Emissions from Stationary Refrigeration and Air Conditioning Systems, Adopted June 7, 1991;

SCAQMD Proposed Amended Rule 1415: Reduction of Refrigerant Emissions from Stationary Refrigeration and Air Conditioning Systems, Version date: April 12, 1994.

"SCAQMD Proposes Amendments to Rule 1415," H. Gwinn Henry, P.E., H.G. Henry and Associates.

"CFC Supply Questions—No Easy Answers," Edward Sullivan, Managing Editor, *Building Operating Management,* April, 1994.

"Refrigerants," Air Conditioning, *Heating & Refrigeration News,* April 11, 1994.

"Survey results show more faith in alternatives, but less in recovery," Melissa Ringwood; B. Checket-Hanks, *Air Conditioning, Heating and Refrigeration News,* April 25, 1994.

"EPA Sees 'Very Tight Supplies' of CFCs as Phaseout Approaches," Thomas A. Mahoney, *The Air Conditioning, Heating and Refrigeration News,* April 25, 1994.

"Selection criteria for long-term refrigerants," Earl B. Muir, Copeland Corp., *The Air Conditioning, Heating and Refrigeration News*, April 4, 1994.

"Are you ready for a Level 2 EPA inspection?," Gene Church Schulz, *The Air Conditioning, Heating and Refrigeration News*, April 4, 1994.

" 'Castrol Retrofill' receives patent for CFC retrofits, *The Air Conditioning, Heating and Refrigeration News*, March 26, 1994.

"EPA Proposes Refrigerant Substitutes for SNAP Program," Charles R. Miró, Issues Manager, *ASHRAE Insights*, Volume 8, number 7, July, 1993.

"From the School of Refrigerant Management," Edward Sullivan, Managing editor, *Building Operating Management*, August, 1993.

"What Contractors should do when EPA comes knocking at their doors," Steven John Fellman, *The Air Conditioning, Heating and Refrigeration News*, September 27, 1993.

"Handbook for Responsible Refrigerant Management, First Edition," Carrier Corporation, 1993.

"Alternative refrigerants for R-22" Special Chiller Section article by Mark W. Spatz, P.E., Member ASHRAE. *ASHRAE Journal*, May 1993.

"CFC phaseout management in all sizes," Paul L. Doppel, *The Air Conditioning, Heating and Refrigeration News*, July 19, 1993.

"State-of-the-Art Centrifugal Chiller Development," Floyd C. Hayes, The Trane Company.

"Effective Refrigerant Management Planning," EPA Guide To CFC Planning, United States Environmental Protection Agency.

"New Guideline to Hold the Line on Recycled Refrigerant Purity," Thomas A. Mahoney, *Business News Publishing Co.*

"CFC Update," published by the Electric Power Research Institute (EPRI), Commercial Building Air-Conditioning Center (CBAC), Issue 5, March 1993.

"Planning makes refrigeration conversions less painful," Eugene Smithart, P.E., The Trane Company, *The Air Conditioning, Heating and Refrigeration News,* April 16, 1993.

"Planning the move to alternatives," Anne M. Hayner, *Engineered Systems,* April, 1993.

"Answers to 10 Key Building Owner Questions about CFCs," The Air-Conditioning and Refrigeration Institute (ARI), *Building Operating Management,* December, 1993.

"R-123: A balanced selection for low pressure systems," Eugene L. Smithart and James G. Crawford, *ASHRAE Journal,* May 1993.

"A contractor's guide: Complying with EPA's refrigerant recycling rule," *Air Conditioning, Heating and Refrigeration News,* May 3, 1993.

Index